Annals of Mathematics Studies

Number 92

THE CLASSIFYING SPACES FOR SURGERY AND COBORDISM OF MANIFOLDS

BY

IB MADSEN AND R. JAMES MILGRAM

PRINCETON UNIVERSITY PRESS
AND
UNIVERSITY OF TOKYO PRESS

PRINCETON, NEW JERSEY
1979

Published in Japan exclusively by
University of Tokyo Press
in other parts of the world by
Princeton University Press

Printed in the United States of America
by Princeton University Press, Princeton, New Jersey

Library of Congress Cataloging in Publication data will
be found on the last printed page of this book

CONTENTS

INTRODUCTION

CHAPTER 1. CLASSIFYING SPACES AND COBORDISM

A. *Bundles with fiber* F *and structure group* Π 3
B. *The classifying spaces for the classical Lie groups* 9
C. *The cobordism classification of closed manifolds* 14
D. *Oriented cobordism theories and localization* 21
E. *Connections between cobordism and characteristic classes* 24

CHAPTER 2. THE SURGERY CLASSIFICATION OF MANIFOLDS

A. *Poincare duality spaces and the Spivak normal bundle* 29
B. *The Browder-Novikov theorems and degree* 1 *normal maps* 34
C. *The number of manifolds in a homotopy type* 38

CHAPTER 3. THE SPACES SG AND BSG

A. *The spaces of stable homotopy equivalences* 45
B. *The space* $Q(S^0)$ *and its structure* 48
C. *Wreath products, transfer, and the Sylow* 2*-subgroups of* Σ_n 51
D. *A detecting family for the Sylow* 2*-subgroups of* Σ_n 54
E. *The image of* $H^*(B\Sigma_n)$ *in the cohomology of the detecting groups* 57
F. *The homology of* $Q(S^0)$ *and* SG 64
G. *The proof of Theorem* 3.32 70

CHAPTER 4. THE HOMOTOPY STRUCTURE OF G/PL AND G/TOP

A. *The* 2*-local homotopy type of* G/PL 77
B. *Ring spectra, orientations and* K*-theory at odd primes* 81
C. *Piece-wise linear Pontrjagin classes* 86
D. *The homotopy type of* G/PL[½] 89
E. *The* H*-space structure of* G/PL 93

CHAPTER 5. THE HOMOTOPY STRUCTURE OF MSPL[½] AND MSTOP[½]

A. *The* KO-*orientation of* PL-*bundles away from* 2 — 99
B. *The splitting of* p-*local* PL-*bundles,* p *odd* — 102
C. *The homotopy types of* G/O[p] *and* SG[p] — 106
D. *The splitting of* MSPL[p], p *odd* — 113
E. *Brumfiel's results* — 116
F. *The map* $f: SG[p] \to BU^{\otimes}[p]$ — 118

CHAPTER 6. INFINITE LOOP SPACES AND THEIR HOMOLOGY OPERATIONS

A. *Homology operations* — 125
B. *Homology operations in* $H_*(Q(S^0))$ *and* $H_*(SG)$ — 130
C. *The Pontrjagin ring* $H_*(BSG)$ — 137

CHAPTER 7. THE 2-LOCAL STRUCTURE OF B(G/TOP)

A. *Products of Eilenberg-MacLane spaces and operations in* $H_*(G/TOP)$ — 142
B. *Massey products in infinite loop spaces* — 149
C. *The proof of Theorem* 7.1 — 154

CHAPTER 8. THE TORSION FREE STRUCTURE OF THE ORIENTED COBORDISM RINGS

A. *The map* $\eta: \Omega_*(G/PL) \to \Omega_*^{PL}$ — 158
B. *The Kervaire and Milnor manifolds* — 164
C. *Constructing the exotic complex projective spaces* — 167

CHAPTER 9. THE TORSION FREE COHOMOLOGY OF G/TOP AND G/PL

A. *An important Hopf algebra* — 174
B. *The Hopf algebras* $F^*(BSO^{\otimes})$ *and* $F^*(G/PL) \otimes \mathbf{Z}[½]$ — 180
C. *The* 2-*local and integral structure of* $F^*(G/PL)$ *and* $F^*(G/TOP)$ — 188

CHAPTER 10. THE TORSION FREE COHOMOLOGY OF BTOP AND BPL

A. *The map* $j_*: F_*(BO) \otimes F_*(G/TOP) \to F_*(BTOP)$ — 193
B. *The embedding of* $F^*(BTOP; \mathbf{Z}_{(2)})$ *in* $H^*(BTOP; Q)$ — 200
C. *The structure of* $\Omega_*^{PL}/\mathrm{Tor} \otimes \mathbf{Z}_{(2)}$ — 205

CHAPTER 11. INTEGRALITY THEOREMS

A. *The inclusion* $F_*(BTOP; \mathbf{Z}[½]) \subset H^*(BTOP; Q)$ — 209
B. *Piece-wise linear Hattori-Stong theorems* — 216
C. *Milnor's criteria for* PL *manifolds* — 221

CHAPTER 12. THE SMOOTH SURGERY CLASSES AND $H_*(BTOP; \mathbf{Z}/2)$

A. *The map* $B(r\times s): B(G/O) \to B^2O \times B(G/TOP)$ 223
B. *The Leray-Serre spectral sequence for* BTOP 231

CHAPTER 13. THE BOCKSTEIN SPECTRAL SEQUENCE FOR BTOP

A. *The Bockstein spectral sequences for* BO, G/TOP *and* B(G/O) 235
B. *The spectral sequence for* BTOP 238
C. *The differentials in the subsequence* 13.21 243

CHAPTER 14. THE TYPES OF TORSION GENERATORS

A. *Torsion generators, suspension and the map* η 246
B. *Torsion coming from relations involving the Milnor manifolds* 249
C. *Applications to the structure of the unoriented bordism rings* $\mathfrak{N}_*^{PD}$ *and* $\mathfrak{N}_*^{PL}$ 251
D. p-*torsion in* Ω_*^{PL} *for* p *odd* 253

APPENDIX. THE PROOFS OF 13.12, 13.13, AND 13.15 256

BIBLIOGRAPHY 263

INDEX 274

INTRODUCTION

In this book we discuss classification results for piecewise linear and topological manifolds. These topics have formed one of the main lines of development for the past 2 decades in the area of algebraic and geometric topology, and recently many of the major problems here have been essentially solved.

Initially our object was to present results on the structure of the piecewise linear and topological bordism rings. However, the need for a fairly comprehensive introduction to the basic algebraic topological results in the theory forced us to expand the discussion considerably. Now the first 8 chapters (the main part of the work) present the homotopy theory of the 'surgery classifying spaces', and the classifying spaces for the various required bundle theories.

The modern development started with the work of Kervaire and Milnor on surgery and the observation of J. H. C. Whitehead on the probable importance of the structure of the normal bundle in classifying structures on manifolds. The Browder-Novikov theorem applied the surgery techniques to show how to classify all PL manifolds in a given homotopy type, and showed that if we have a (simply connected) space X which satisfies the homology conditions for an n dimensional closed manifold (Poincaré duality) with $n \geq 5$, then there is a PL manifold M in the homotopy type of X if and only if a certain fibering $Y \to X$ with fiber having the homotopy type of a sphere can be replaced by a piecewise linear fiber bundle.

This pivotal result reduced the basic questions to questions involving classifying homotopy sphere bundles, finding effective ways of telling when they reduce to PL bundles, and then (for estimating the number of manifolds in a homotopy type) counting the number of distinct equivalence classes of PL bundles which correspond to the original homotopy sphere bundle.

A theorem of Stasheff identifies the equivalence class of a sphere bundle $Y \to X$ with a homotopy class of maps

$$\phi : X \to BG$$

where BG serves as a classifying space for spherical fiberings, fixed and independent of X. Moreover, there is a similar classifying space for PL-sphere bundles BPL, a map $p: BPL \to BG$ so that PL structures on the given sphere bundle correspond to homotopy classes of commutative diagrams

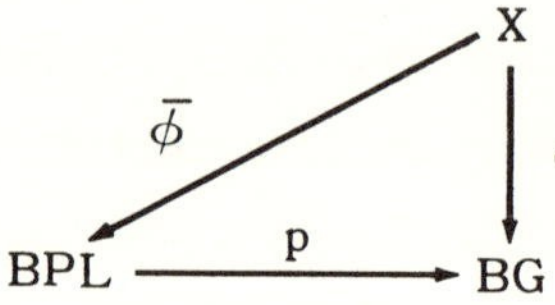

It is customary to denote the homotopy fiber of the map p as G/PL and the equivalence classes of diagrams above are in one-one correspondence with the set of homotopy classes of maps of X into G/PL.

Thus, the basic problems were reduced to problems in homotopy theory, and attention turned to the study of these spaces, as well as the maps between them.

By the mid 1960's it was realized (by Sullivan and others) that the results of Kervaire and Milnor showed that in dimension ≥ 5

$$\pi_i(G/PL) = \begin{cases} Z & i \equiv 0(4) \\ 0 & i \equiv 1(2) \\ Z/2 & i \equiv 2(4) \end{cases}$$

and results of Cerf and an exact sequence of Hirsch and Mazur then extended this to all i.

D. Sullivan attacked and solved the problem of determining the homotopy type of G/PL —introduced ideas of localization into the theory—and obtained much information on the homotopy type of BPL when localized at odd primes.

Milgram at this point began studying the cohomology groups of BG after work of Milnor and Gitler-Stasheff had given some low dimensional information, and he obtained $H^*(BG; Z/2)$.

Madsen then examined the infinite loop space structure of G/PL and BG, evaluated their homology operations and described the Pontrjagin rings $H_*(BG; Z/2^i)$. J. P. May and A. Tsuchiya gave complete information for $H_*(BG; Z/p)$ and May later studied $H_*(BG; Z/p^i)$, when p is odd.

At this point results of Kirby, Kirby-Siebenmann and Lashof-Rothenberg showed how to include topological manifolds into the picture, by providing an analogous theory with the space BTOP classifying topological sphere fiberings, the map $p': BTOP \to BG$ and the homotopy fiber of p', G/TOP.

Kirby Siebenmann showed that the natural map $G/PL \to G/TOP$ has fiber the Eilenberg-MacLane space $K(Z/2, 3)$ and then obtained the homotopy type of G/TOP also. (See [65] Annex B.)

At this point Sullivan's results, together with those of May and Tsuchiya gave a fairly good cohomological picture of the story at odd primes, but not much was known at the prime 2.

In 1970 Brumfiel, Madsen and Milgram began the study of the map $i: G \to G/PL$ at the prime 2. The induced map on $Z/2$ cohomology was calculated and these results later enabled Madsen and Milgram to complete the study of i at the prime 2.

In particular, these results, combined with the results at odd primes give effective methods for deciding when a homotopy sphere bundle $Y \to X$ can be replaced by a piecewise linear or topological bundle, as well as allowing the determination of a large part of the piece-wise linear and topological bordism rings.

Here we present the theory outlined above and determine the torsion free parts of the oriented bordism rings Ω_*^{TOP}/Torsion ($= \Omega_*^{PL}$/Torsion). We give an explicit set of generators: the differentiable generators, a set of exotic projective spaces, and the "Milnor manifolds" of index 8. We also obtain integrality theorems for characteristic classes on piecewise linear and topological manifolds extending the now classical Riemann-Roch theorems for differential manifolds.

Much of the material covered in this book appears in print with detailed proofs for essentially the first time. In Chapter 3 we give a direct method for calculating the homology of G and BG at the prime 2. This somewhat non-standard treatment owes much to discussions with Ben Mann. In Chapters 4 and 5 we review results mainly due to Sullivan on G/PL and MSPL and their topological analogues. In particular the results in Chapter 5 have no complete proofs in the literature. Moreover, our proofs seem somewhat simpler than the proofs outlined by Sullivan in [135]. Chapter 6 is a very brief introduction to the theory of infinite loop spaces and homology operations, and in Chapter 7 we prove the result, basic to our theory, that B(G/TOP) when localized at 2 is a product of Eilenberg-MacLane spaces. This implies that the 2-local part of the obstruction to reducing a spherical fibering to a PL bundle is purely a characteristic class obstruction.

In the later chapters we put these results together with some rather unpleasant calculations to study the integral cohomology of BPL and BTOP. These results answer the questions of exactly which polynomials in the Pontrjagin classes are integral topological invariants, and are applied to give fairly complete information on the torsion free parts of the piecewise linear and topological oriented bordism rings.

Finally we would like to thank Erkki Laitinen for his helpful comments and for preparing the index, and above all Greg Brumfiel for vital suggestions and help when we were mired in difficulties.

IB MADSEN

R. JAMES MILGRAM

The Classifying Spaces
for Surgery and Cobordism
of Manifolds

CHAPTER 1

CLASSIFYING SPACES AND COBORDISM

A basic technique in topology is to reduce a geometric classification problem to a homotopy classification of maps into an associated "classifying space" or universal object. In this chapter we review the classifying spaces for various bundle theories and the connection between the cobordism classification of manifolds and the homotopy groups of the associated Thom spectra.

A. *Bundles with fiber* F *and structure group* Π

Suppose we are given a topological group Π and a fixed action of Π on a space F. By an (F,Π)-bundle we shall mean a locally trivial bundle with fiber F and structure group Π (cf. Steenrod [129]). A principal Π-bundle is an (F,Π)-bundle with $F = \Pi$ and with the Π-action given as the usual product in Π.

DEFINITION 1.1. A space B is said to be a classifying space for (F,Π)-bundles if there is an (F,Π)-bundle E over B such that for any (F,Π)-bundle $\tilde{E}$ over a finite dimensional CW complex X there is a unique homotopy class of maps $f\colon X \to B$ with $f^*(E) \cong \tilde{E}$.

In this definition one may suppress F since associated to any (F,Π)-bundle E there is a principal Π-bundle $\mathrm{Prin}_\Pi(E)$ which determines E, $E = \mathrm{Prin}_\Pi(E) \times_\Pi F$. Indeed $\mathrm{Prin}_\Pi(E)$ is the space of (F,Π)-bundle maps

$$\begin{array}{ccc} F & \longrightarrow & E \\ \big\downarrow & & \big\downarrow \\ * & \longrightarrow & X \end{array}$$

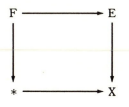

Hence B classifies (F,Π)-bundles if and only if it classifies principal Π-bundles.

In the generality of 1.1 there is no reason to assume B is a CW complex, but if B_1 is the geometric realization of the singular complex of B and $p: B_1 \to B$ the natural map then $p_*: [X, B_1] \to [X, B]$ is a bijection for any finite dimensional CW complex X. Moreover, two CW complexes B_1 and B_2 which both classify principal Π-bundles are homotopy equivalent. (To see this let $B_1^{(i)}$ be the i-skeleton and consider the unique homotopy class of maps $k_i: B_1^{(i)} \to B_2$ classifying the restriction of E_1 to $B_1^{(i)}$. The homotopy extension theorem guarantees an (in general non-unique) map $k: B_1 \to B_2$ with $k|B_1^{(i)} \simeq k_i$ for all i. Similarly, we get a map $\ell: B_2 \to B_1$ and the compositions $k \circ \ell$ and $\ell \circ k$ restrict to maps homotopic to the inclusions on the i-skeleta. In particular B_1 and B_2 are weakly homotopy equivalent hence homotopy equivalent.)

It is standard to denote by $B\Pi$ any classifying space for principal Π-bundles and similarly write $E\Pi$ for the Π-bundle over $B\Pi$. The spaces $B\Pi$, $E\Pi$ exist for any topological group Π. The original construction is due to Milnor [98] and is based on the following "recognition principle" [129].

THEOREM 1.2 (Steenrod). *Let* E *be a principal* Π*-bundle over* B. *Then* B *is a classifying space for principal* Π*-bundles if and only if* E *is connected and* $\pi_i(E) = 0$ *for* $i > 0$.

Alternatively, the existence of $B\Pi$ follows rather easily from Brown's representation theorem (see e.g. [30], [45]).

More generally, Dold and Lashof constructed a principal quasi-fibering (see [48])

$$\Pi \to E\Pi \to B\Pi$$

when Π is a topological monoid which is either path connected or whose set of path components forms a group, and Stasheff [128] proved (see also [45])

THEOREM 1.3 (Stasheff). *Suppose* F *is a finite* CW *complex and let* $H(F)$ *be the monoid of homotopy equivalences* $f: F \to F$ *equipped with the compact-open topology. Then for a finite dimensional* CW *complex* X, $[X, BH(F)]$ *is the set of equivalence classes of homotopy* F*-bundles over* X.

Here a homotopy F-bundle over X is a map $f: Y \to X$ so that when we convert f into a Serre-fibering the fiber has the homotopy type of F, and two such bundles are equivalent if there is a homotopy equivalence $h: Y \to Y'$ with $f'h = f$,

$$\begin{array}{ccc} Y & \xrightarrow{\ h\ } & Y' \\ \big\downarrow f & & \big\downarrow f' \\ X & \xrightarrow{\ Id\ } & X \end{array}$$

We now give an alternative construction of the classifying space $B\Pi$ which has certain formal advantages over those above (cf. [119], [131]). We will assume that Π is a topological monoid or group with identity e and we require that e be a neighborhood deformation retract (N.D.R.) in Π. (Specifically this means there is a function $f: \Pi \to [0,1]$ with $f^{-1}(0) = e$, $f^{-1}[0,1) = N$ and a deformation $H: \Pi \times [0,1] \to \Pi$ with $H(N \times [0,1)) \subset N, H(n,1) = n$, $H(n,0) = e$ and $fH(n,t) = tf(n)$ for $n \in N$.) This condition is satisfied by any Lie group or a CW complex. It is trivially satisfied if Π is discrete.

Let σ^n be the standard n-simplex coordinatized as the set of points $(t_1, \cdots, t_n) \in I^n$ with $1 \geq t_1 \geq t_2 \geq \cdots \geq t_n \geq 0$. Its i'th face $\tilde{\partial}_i \sigma^n$ is then specified by

$$\begin{array}{lll} t_1 = 1 & \text{if} & i = 0 \\ t_i = t_{i+1} & \text{if} & 0 < i < n \\ t_n = 0 & \text{if} & i = n \end{array}$$

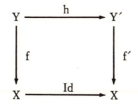

Now, set

$$E^0\Pi = \coprod_{n \geq 0} \sigma^n \times \Pi^n \times \Pi$$

where each $\sigma^n \times \Pi^n \times \Pi$ is given the compactly generated topology. (That is to say U is open if and only if its intersection with each compact set K of $\sigma^n \times \Pi^n \times \Pi$ is open in the induced topology on K.)

We define an equivalence relation in $E^0\Pi$ by

$$(t_1, \cdots, t_n, g_1, \cdots, g_n, g) \sim (t_2, \cdots, t_n, g_2, \cdots, g_n, g) \quad \text{for} \quad t_1 = 1$$

1.4 $$(t_1, \cdots, t_n, g_1, \cdots, g_n, g) \sim (t_1, \cdots, \hat{t}_i, \cdots, t_n, g_1, \cdots, g_{i-1}g_i, \cdots, g_n, g)$$
$$\text{for } t_{i-1} = t_i \quad \text{or} \quad g_i = e$$

$$(t_1, \cdots, t_n, g_1, \cdots, g_n, g) \sim (t_1, \cdots, t_{n-1}, g_1, \cdots, g_{n-1}, g_n g) \quad \text{for} \quad t_n = 0$$

We give $E\Pi = E^0\Pi/\sim$ *) the quotient topology which is also compactly generated. In particular, there is a compactly continuous function

$$E\Pi \times \Pi \to E\Pi$$

coming from the action of Π on $E^0\Pi$ defined on points by

$$(t_1, \cdots, t_n, g_1, \cdots, g_n, g)\, g' = (t_1, \cdots, t_n, g_1, \cdots, g_n, gg') .$$

We set $B\Pi$ equal to the orbit space of this action. Alternately, $B\Pi$ can be described in terms of an equivalence relation similar to 1.4 on $B^0\Pi = \coprod_{n \geq 0} \sigma^n \times \Pi^n$.

The following theorem is proved in [131]. ($e \in \Pi$ is supposed to be an N.D.R.)

---

*) As a categorical construction the space $E\Pi$ can be regarded as the balanced product (tensor product) of the unreduced bar construction of Π ([80], p. 248) and the category $\boldsymbol{\sigma}$ whose objects are the σ^n and whose morphisms are (generated by) the face maps $\tilde{\partial}_i$ and the degeneracy maps $\tilde{s}_i : \sigma^n \to \sigma^{n-1}$ which omit the coordinate t_i.

THEOREM 1.5. (a) *If* Π *is a topological group then* $E\Pi \to B\Pi$ *is a principal* Π*-bundle.*

(b) *If* Π *is a topological monoid then* $E\Pi \to B\Pi$ *is a principal* Π *quasi-fibering,*

(c) $E\Pi$ *is contractible,*

(d) $B(\Pi_1 \times \Pi_2) = B\Pi_1 \times B\Pi_2$,

when the cartesian products are given the compactly generated topology.

(e) *The construction is natural in the sense that if* $f: \Pi \to \Gamma$ *is a continuous homomorphism then there are induced maps* $Ef: E\Pi \to E\Gamma$ *and* $Bf: B\Pi \to B\Gamma$ *satisfying the usual functorial properties.**)

The contraction in (c) comes from $H_t: E^0\Pi \to E^0\Pi$ defined by

$$H_t(t_1, \cdots, t_n, g_1, \cdots, g_n, g) = (\overline{t_1+t}, \cdots, \overline{t_n+t}, t, g_1, \cdots, g_n, g, e)$$

where $\bar{t} = t$ if $t \le 1$ and $\bar{t} = 1$ if $t \ge 1$. Property (d) is proved by checking directly that

$$B\pi_1 \times B\pi_2 : B(\Pi_1 \times \Pi_2) \to B\Pi_1 \times B\Pi_2$$

is a homeomorphism, where π_i is the projection on Π_i, and of course

$$Bf(t_1, \cdots, t_n, g_1, \cdots, g_n) = (t_1, \cdots, t_n, f(g_1), \cdots, f(g_n)) .$$

The more difficult part of the proof is establishing (a) and (b). This is where the N.D.R. assumption on e comes in.

---

*)Among the natural properties of $B\Pi$ is a filtration ($B\Pi^{(n)}$ = image of $\coprod_{k\le n} \sigma^n \times \Pi^n$ in $B\Pi$). Filtrations give rise to spectral sequences and thus one has a natural spectral sequence associated with $B\Pi$. One can check that with field coefficients k, its E^1-term is the "bar construction" on $H_*(\Pi; k)$. Hence its E^2-term is $\mathrm{Tor}_{H_*(\Pi)}(k, k)$ and

$$\mathrm{Tor}_{H_*(\Pi)}(k, k) \Longrightarrow H_*(B\Pi; k) .$$

In cohomology we have dually

$$\mathrm{Ext}_{H_*(\Pi)}(k, k) \Longrightarrow H^*(B\Pi; k) .$$

These are the Eilenberg-Moore spectral sequences [115], [151] and are extremely powerful computational tools.

EXAMPLE 1.6. Let Π be a discrete group; then $E\Pi \to B\Pi$ is a principal Π-bundle. Since $E\Pi$ is contractible, $E\Pi$ is the universal covering space of $B\Pi$ and $\pi_1(B\Pi) = \Pi$, $\pi_i(B\Pi) = 0$ for $i > 1$. Moreover, $B\Pi$ is a CW complex with cells of the form $[g_1|\cdots|g_n]$ ($g_i \neq e$ for all i) corresponding to $\sigma^n \times (g_1, \cdots, g_n)$. The cellular "boundary relation" is

$$\partial[g_1|\cdots|g_n] = [g_2|\cdots|g_n] + \sum_{i=1}^{n-1} (-1)^i[g_1|\cdots|g_ig_{i+1}|\cdots|g_n] + (-1)^n[g_1|\cdots|g_{n-1}].$$

EXAMPLE 1.7. Let Π be abelian so that the multiplication $\mu: \Pi \times \Pi \to \Pi$ is a homomorphism. Using 1.5(d) we have

$$B\mu: B\Pi \times B\Pi \to B\Pi$$

which makes $B\Pi$ into an associative abelian H-space with unit. Moreover, $B\Pi$ is a CW complex so it satisfies the N.D.R. property for the unit and we can iterate the construction. In particular, if Π is a discrete abelian group, then $K(\Pi, 1) = B\Pi$ is an abelian topological group,

$$(t_1, \cdots, t_n, g_1, \cdots, g_n)^{-1} = (t_1, \cdots, t_n, g_1^{-1}, \cdots, g_n^{-1})$$

and $BK(\Pi, 1) = K(\Pi, 2)$ is again such so we can construct all the Eilenberg-MacLane spaces by iteration.

EXAMPLE 1.8. The classical Lie groups U_n, O_n and SO_n give classifying spaces: BU_n classifying principal U_n-bundles or equivalently complex vector bundles, BO_n classifying principal O_n bundles or real vector bundles, and BSO_n classifying oriented real vector bundles. These spaces have been extensively studied. We need certain facts about them which we will recall after the next two examples. For now, note that $U_1 = S^1 = K(Z, 1)$ so $BU_1 = K(Z, 2) = CP^\infty$ and $O_1 = Z/2$ so $BO_1 = K(Z/2, 1) = RP^\infty$.

EXAMPLE 1.9. Let TOP_n be the topological monoid of homeomorphisms $f: R^n \to R^n$, $f(0) = 0$, with the compact-open topology and let G_n denote the set of homotopy equivalences of S^{n-1} also with the compact-open topology. Then $B\,TOP_n$ and BG_n classify R^n-bundles with a zero-section and homotopy S^{n-1}-bundles (*also called spherical fibrations*), respectively.

EXAMPLE 1.10. For the piecewise linear homeomorphisms of $\mathbf{R}^n$ one has to proceed a little differently: the compact-open topology is not suitable. Instead we let PL_n denote the simplicial group whose k-simplices consist of all piecewise linear homeomorphisms $f: \Delta^k \times \mathbf{R}^n \to \Delta^k \times \mathbf{R}^n$ commuting with the projection on Δ^k and preserving the zero section. The geometric realization of PL_n again denoted PL_n is a topological monoid whose classifying space BPL_n classifies piecewise linear $\mathbf{R}^n$-bundles ([68], [69]).

B. *The classifying spaces for the classical Lie groups*

At several later points in the book we will need some more or less standard results on the spaces BU_n, BO_n which we collect here for convenience. Most of these results can be found in one form or another in e.g. [54], [102].

To begin there are Whitney sum maps

1.11
$$\phi_{n,m}: BU_n \times BU_m \to BU_{n+m}$$
$$\phi_{n,m}: BO_n \times BO_m \to BO_{n+m}$$

induced from the homomorphisms

$$\phi_{n,m}: U_n \times U_m \to U_{n+m}, \quad \phi_{n,m}: O_n \times O_m \to O_{n+m}$$

given by

$$\phi_{n,m}(A, B) = \begin{pmatrix} A & 0 \\ 0 & B \end{pmatrix}.$$

Iterating this we obtain maps

$$B\phi_n : BU_1 \times \cdots \times BU_1 \to BU_n$$

$$B\phi_n : BO_1 \times \cdots \times BO_1 \to BO_n .$$

Now, $BU_1 = K(\mathbf{Z}, 2) \simeq CP^\infty$ has cohomology ring $H^*(CP^\infty; \mathbf{Z}) = P\{d\}$, the polynomial algebra on a 2-dimensional generator. Similarly, $O_1 = \mathbf{Z}/2$ and $BO_1 = K(\mathbf{Z}/2, 1) \simeq RP^\infty$ has cohomology ring $H^*(RP^\infty; \mathbf{Z}/2) = P\{f\}$, where f is 1-dimensional.

Let $T: U_1 \times U_1 \to U_1 \times U_1$ be the interchange map, then $\phi_2 \circ T \simeq \phi_2$ since setting $H_t = R(\frac{1}{2}\pi t)^{-1} \cdot \phi_2 \cdot R(\frac{1}{2}\pi t)$ where

$$R(\tfrac{1}{2}\pi t) = \begin{pmatrix} \cos(\frac{1}{2}\pi t) & \sin(\frac{1}{2}\pi t) \\ -\sin(\frac{1}{2}\pi t) & \cos(\frac{1}{2}\pi t) \end{pmatrix}$$

gives a 1-parameter group of homomorphisms $U_1 \times U_1 \to U_2$ starting at $t = 0$ with ϕ_2 and ending with $\phi_2 \circ T$ at $t = 1$. By the naturality property 1.5(e) this gives a continuous 1-parameter family of maps $BU_1 \times BU_1 \to BU_2$ starting with $B\phi_2$ and ending with $(B\phi_2) \circ T$. Iterating this, if $\lambda : BU_1 \times \cdots \times BU_1 \to BU_1 \times \cdots \times BU_1$ is any coordinate permutation then $(B\phi_n) \circ \lambda \simeq B\phi_n$. (Similarly for O_n.)

Thus the image $(B\phi_n)^*(H^*(BU_n; \mathbf{Z}))$ is invariant under the action of the symmetric group Σ_n, and conversely we have

THEOREM 1.12. (a) $B\phi^*$ *injects* $H^*(BU_n; \mathbf{Z})$ *into* $P\{d_1, \cdots, d_n\}$ *as the subalgebra of invariants under the symmetric group* Σ_n.

(b) $B\phi^*$ *injects* $H^*(BO_n; \mathbf{Z}/2)$ *into* $P\{f_1, \cdots, f_n\}$ *as the subalgebra of invariants under* Σ_n.

For a proof see e.g. [102]. Usually this result is proved by first constructing the Chern classes and Stiefel-Whitney classes. But a proof based on the generalized transfer of Becker and Gottlieb can also be given [34].

DEFINITION 1.13. The i'th Chern class $c_i \in H^{2i}(BU_n; \mathbf{Z})$ is the class so that $B\phi^*(c_i) = \sigma_i(d_1, \cdots, d_n)$. The i'th Stiefel-Whitney class in

$H^i(BO_n; Z/2)$ is the class so that $B\phi^*(w_i) = \sigma_i(f_1, \cdots, f_n)$. (Here σ_i denotes the i'th elementary symmetric polynomial in the stated variables.)*)

From 1.12 we then have

$$H^*(BU_n; Z) = P\{c_1, \cdots, c_n\}$$
$$H^*(BO_n; Z/2) = P\{w_1, \cdots, w_n\}$$

The Whitney sum maps in 1.11 induce homomorphisms

$$B\phi^*_{n,m} : H^*(BU_{n+m}) \to H^*(BU_n) \otimes H^*(BU_m)$$
$$B\phi^*_{n,m} : H^*(BO_{n+m}) \to H^*(BO_n) \otimes H^*(BO_m)$$

given by

1.14 $$B\phi^*_{n,m}(c_i) = \sum_{j=0}^{i} c_j \otimes c_{i-j}, \quad B\phi^*_{n,m}(w_i) = \sum_{j=0}^{i} w_j \otimes w_{i-j},$$

where we take $c_j \in H^*(BU_n)$ and $w_j \in H^*(BO_n)$ to be zero if $j > n$. This all follows from 1.12 by simple diagram chase. Note in particular that if $Bi: BU_n \to BU_{n+1}$ is induced from the inclusion then

$$Bi^*(c_k) = c_k \quad \text{for} \quad k \le n \quad \text{and} \quad Bi^*(c_{n+1}) = 0$$

since Bi can be considered as the composition

$$BU_n \times (*) \to BU_n \times BU_1 \xrightarrow{B\phi} BU_{n+1}.$$ Similarly for $Bi: BO_n \to BO_{n+1}$.

REMARK 1.15. It is worth noting that BO_n has the homotopy type of the universal S^n bundle over BO_{n+1}, and there is a homotopy equivalence

$$f: BO_n \to \text{Universal } S^n\text{-bundle}$$

so that $\pi f \simeq Bi: BO_n \to BO_{n+1}$.

Indeed the sphere bundle is given by

---

*)It is a classical theorem that $P\{x_1, \cdots, x_n\}^{\Sigma_n} = P\{\sigma_1, \cdots, \sigma_n\}$ for field or integer coefficients.

$$\pi : EO_{n+1} \times_{O_{n+1}} S^n \to BO_{n+1} .$$

This is the universal (S^n, O_{n+1})-bundle and a mapping $X \to EO_{n+1} \times_{O_{n+1}} S^n$ is equivalent to specifying an S^n-bundle over X along with a section. Hence the associated R^{n+1} bundle is the Whitney sum of an R^n-bundle and a trivial line bundle. This gives a map

$$EO_{n+1} \times_{O_{n+1}} S^n \to BO_n$$

which is a homotopy equivalence. Moreover, if we identify $EO_{n+1} \times_{O_{n+1}} S^n$ with BO_n then π is identified with i. Similar remarks apply to $BU_n \to BU_{n+1}$ and $BSO_n \to BSO_{n+1}$.

Consider the disjoint unions $W(U) = \coprod_{n \geq 0} BU_n$, $W(O) = \coprod_{n \geq 0} BO_n$. The map $\amalg B\phi_{n,m}$ make these spaces into associative H-spaces with unit $* = BO_0 = BU_0$. This makes $H_*(W)$ into an associative ring with unit, and dualizing 1.12 we have

$$\begin{aligned} H_*(W(U); Z) &= P\{b_2, \cdots, b_{2n}, \cdots\} \times Z^+ \\ H_*(W(O); Z/2) &= P\{e_1, \cdots, e_n, \cdots\} \times Z^+ \end{aligned}$$

where $b_{2n} \in H_{2n}(BU_1; Z)$ is dual to d^n and $e_n \in H_n(BO_1; Z/2)$ is the non-zero element. The "group completions"*) of W(U) and W(O) are the classifying spaces $BU \times Z$ and $BO \times Z$ for complex and real K-theory, classifying virtual vector bundles ($BU = \varinjlim BU_n$ and $BO = \varinjlim BO_n$). In particular

1.16
$$\begin{aligned} H_*(BU; Z) &= P\{b_2, b_4, \cdots\} \\ H_*(BO; Z/2) &= P\{e_1, e_2, \cdots\} \end{aligned}$$

We have inclusions $r : U_n \to O_{2n}$ and $c : O_n \to U_n$ associated re-

---

*) This is a fundamental notion introduced by Barratt [15], Barratt-Priddy [16] and D. Quillen [112]. There are two approaches. It is easily checked that the spaces W( ) above are free associative unitary monoids. Such an object embeds uniquely into a minimal topological group—its group completion. Alternatively, the group completions can be defined as the loop space $\Omega BW(\ )$ (see also [120]).

spectively with realification which regards C^n as R^{2n} and complexification $R^n \to R^n \otimes C = C^n$. The compositions

$$c \circ r: U_n \to U_{2n}$$
$$r \circ c: O_n \to O_{2n}$$

are conjugate to the inclusions

1.17 $$A \to \begin{pmatrix} A & 0 \\ 0 & \bar{A} \end{pmatrix}, \quad A \to \begin{pmatrix} A & 0 \\ 0 & A \end{pmatrix}.$$

On the classifying space level we obtain maps $c: BO_n \to BU_n$ and $r: BU_n \to BO_{2n}$ and the compositions $c \circ r$ and $c \circ r$ are homotopic to the maps induced from 1.17 (cf. 3.11).

In terms of these maps the Pontrjagin classes p_{4i} are defined by

$$p_{4i} = (-1)^i c^*(c_{2i}) \in H^{4i}(BO; Z)\,.^{*)}$$

Reduced mod 2, $p_{4i} = w_{2i}^2$. The complete cohomology structure of $BO = \varinjlim BO_n$ can now be summarized in

THEOREM 1.18. (a) *The torsion in* $H^*(BO; Z)$ *is a vector space over* $Z/2$ *and*

$$H^*(BO; Z)/\mathrm{Tor} = P\{p_4, \cdots, p_{4i}, \cdots\}\,.$$

(b) *The diagonal in* $H^*(BO; Z)/\mathrm{Tor}$ *induced from the* H-*space structure is given by*

$$\psi(p_{4i}) = \sum_{j=0}^{i} p_{4j} \otimes p_{4(i-j)} \qquad (p_0 = 1)\,.$$

---

*)It is customary to denote the i'th Pontrjagin class by p_i rather than p_{4i}. In this book, however, we find it convenient to always have the subscripts indicate the dimension of the characteristic classes. Indeed only in Chapter 9 are there any exceptions to this rule.

Proof. From 1.12(b) it follows that the Bockstein $\beta: H^*(BO; \mathbf{Z}/2) \to$
$H^*(BO; \mathbf{Z}/2)$ maps w_{2n} to $w_{2n+1} + w_1 w_{2n}$ and w_{2n+1} to $w_1 \cdot w_{2n+1}$.
Since β is a derivation it is easily checked that the homology of the
chain complex $(H^*(BO; \mathbf{Z}/2), \beta)$ is the polynomial algebra
$P\{w_2^2, w_4^2, \cdots, w_{2n}^2, \cdots\}$. But w_{2n}^2 is the reduction of the integral class
p_{4n} by previous remarks. Hence the mod. 2 Bockstein spectral se-
quence ([23]) for BO has $E^2 = E^\infty$ so the 2-torsion consists of elements
of order 2, and

$$H^*(BO; \mathbf{Z})/\mathrm{Tor} \otimes \mathbf{Z}_{(2)} = P\{p_4, \cdots, p_{4i}, \cdots\} .$$

$(\mathbf{Z}_{(2)}$ is the subring of fractions $a/b \in \mathbf{Q}$ with denominator prime to 2.)
On the other hand comparing $H^*(BO; \mathbf{Z})$ with $H^*(BU; \mathbf{Z})$ via realification
and complexification we see that there is no odd-primary torsion in
$H^*(BO; \mathbf{Z})$. This proves (a).

The second part of 1.18 follows from 1.14 upon noting that
$2c^*(c_{2i+1}) = 0$.

Theorem 1.18 remains true if we replace BO with its two fold covering BSO = $\lim_{\to}$ BSO$_n$. The mod. 2 cohomology of BSO is also easy.

$$H^*(BSO_n; \mathbf{Z}/2) = P\{w_2, \cdots, w_n\} .$$

REMARK 1.19. For the spaces H_n of 1.9 and 1.10 we also have
"Whitney sum" maps

$$B\phi_{n,m} : BH_n \times BH_m \to BH_{n+m} .$$

(If $H_n = G_n$ then the Whitney sum is the fiberwise join of spherical fibrations.) The $B\phi_{n,m}$ induce H-structures on the associated stable objects,
BH = $\lim_{\to}$ BH$_n$.

C. *The cobordism classification of closed manifolds*

A large part of our discussion after Chapter 7 will deal with the
cobordism rings of topological and triangulated manifolds. Also, throughout

the next chapters, particularly Chapter 4 and Chapter 5 we will need some basic facts on the cobordism rings for differentiable manifolds. In this section and the next two we collect the facts we will need. Good references containing more details are [40], [102], [133].

Roughly, two closed manifolds M_1^n, M_2^n are cobordant if there is a compact manifold W^{n+1} with $\partial W = M_1 \dot{\cup} M_2$ (disjoint union); but this needs some elaboration. First, by manifold we could mean either smooth manifold, piecewise linear (PL) manifold or topological manifold—the three basic manifold categories. Second, manifolds can come with extra structure, e.g. be oriented, have Spin structure or be weakly complex. Our prime concern is the case of oriented manifolds in any of the three categories. However, to avoid unnecessary repetition we shall simply talk about (H)-manifolds without further specification of the structure (H).

We call two (H)-manifolds equal if they are "isomorphic" (e.g. for oriented PL manifolds this means PL-homeomorphic via an orientation preserving map). For our (H)-manifolds there is a 1-1 correspondence (via restriction) between (H)-structures on $M = M \times 0$ and (H)-structures on $M \times I$. The inverse (H)-structure, denoted $-M$, is the one induced on $M \times 1 \subset M \times I$ when $M \times I$ is given the (H)-structure coming from $M = M \times 0$.

Our (H)-manifolds have an associated (H)-"bundle theory": There is a sequence of topological monoids H_n with BH_n classifying H-bundles (with a zero section), e.g. $H_n = SPL_n$ with BH_n classifying oriented PL R^n-bundles. The universal (n-plane) bundle over BH_n is denoted γ_H^n. If $V^k \subset M^{n+k}$ is an (H)-submanifold and the codimension n is sufficiently large then there is an (H)-normal bundle $\nu^n = \nu^n(V^k : M^{n+k})$ ([96], [66]) and a classifying diagram

$$\begin{array}{ccc} \nu^n & \xrightarrow{\hat{g}} & \gamma_H^n \\ \downarrow & & \downarrow \\ V^k & \xrightarrow{g} & BH_n \end{array}$$

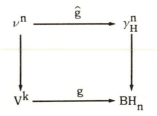

DEFINITION 1.20. Two closed (H)-manifolds M_1^n and M_2^n are called (H)-cobordant if there is an (H)-manifold W^{n+1} with $\partial W = M_1 \dot{\cup} (-M_2)$ such that the (H)-structure of W restricts to those of M_1 and $-M_2$.

The cobordism relation of 1.20 is an equivalence relation and provisionally we denote the set of equivalence classes by T_*^H. It becomes an abelian group when we set $\{M_1^n\} + \{M_2^n\} = \{M_1^n \dot{\cup} M_2^n\}$.

A singular (H)-manifold in a space X is a pair (M, f) consisting of a closed (H)-manifold and a map $f: M \to X$.

DEFINITION 1.21. Two singular (H)-manifolds (M_1^n, f_1) and (M_2^n, f_2) are called cobordant if there is an (H)-manifold W with $\partial W = M_1 \dot{\cup} (-M_2)$ and a map $F: W \to X$ so that $F|M_i = f_i$.

Once more cobordism of pairs (M, f) is an equivalence relation and the resulting equivalence classes form an abelian group $T_*^H(X)$. If (H)-structures on M, N induce a unique (H)-structure on $M \times N$ (which is isomorphic to the (H)-structure on $N \times M$) then $T_*^H = T_*^H(pt)$ becomes an associative, commutative, unitary ring and $T_*^H(X)$ becomes a T_*^H-module: $\{M\} \cdot \{N, f\}$ is the class of $M \times N \xrightarrow{p_2} N \xrightarrow{f} X$.

The final group we wish to construct is $T_*^H(X, Y)$ for a pair of CW-complexes $Y \subset X$. This is given by taking equivalence classes of pairs (W, f),

$$f: (W, \partial W) \to (X, Y)$$

where W is an (H)-manifold. The cobordism relation is given by an (H)-manifold V with $\partial V = W_1 \cup_\partial V_0 \cup_\partial -W_2$ and a map $F: (V, V_0) \to (X, Y)$

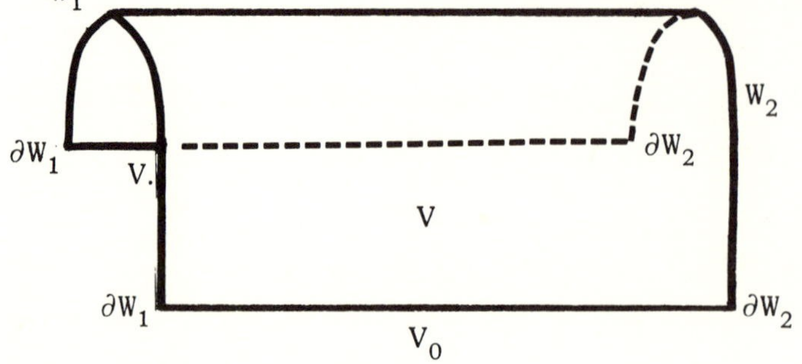

There is a long exact sequence

$$\cdots \longrightarrow T_n^H(Y) \xrightarrow{i_*} T_n^H(X) \xrightarrow{j_*} T_n^H(X,Y) \xrightarrow{\partial} T_{n-1}^H(Y) \longrightarrow \cdots$$

where $\partial\{W, f\} = \{\partial W, f|\partial W\}$ and i_* and j_* are the obvious inclusions. Moreover, given $g: X \to Z$, then g induces a map $g_*: T_*^H(X) \to T_*^H(Z)$ by $g_*\{M, f\} = \{M, gf\}$ and one easily checks that if $g_1 \simeq g_2$, then $g_{1*} = g_{2*}$ so $T_*^H(\ )$ defines an additive homotopy functor.

Next, we examine the excision properties for T_*^H. Write $\tilde{T}_*^H$ for the kernel of $T_*^H(X) \to T_*^H(pt)$ and let

$$\phi: T_*^H(X,Y) \to \tilde{T}_*^H(X \cup cY), \qquad cY = Y \times I / Y \times 1$$

be the map which sends $\{W, f\}$ to $\{W \underset{\partial}{\cup} -W, f \cup g\}$. Here $g: W \to cY \subset X \cup cY$ is an extension of $f|\partial W$, mapping a collar $[0,1] \times \partial W \subset W$ to cY by $Id \times f$ and the rest of W to the cone point. We want to construct an inverse to ϕ,

$$\lambda: \tilde{T}_*^H(X \cup cY) \to T_*^H(X,Y) .$$

For this we need transversality to hold for (H)-structures.

Let Z be a complex, suppose an (H)-bundle ξ^k with base space X is embedded in Z, and let M^{n+k} be an (H)-manifold. Recall that a map $f: M^{n+k} \to Z$ is called transverse to X if $f^{-1}(X) = V^n$ is an (H)-submanifold of M^n with (H)-normal bundle ν^k and if there is an embedding of ν^k into a tubular neighborhood of V^n in M^{n+k} so that $f|\nu^k: \nu^k \to \xi^k$ is an (H)-bundle map.

We say that the (H)-category *satisfies transversality* (of codimension k) if every singular (H)-manifold in Z,

$$f: M^{n+k} \to Z$$

can be deformed slightly to a map transverse to X.

We use this for a singular (H)-manifold $f: M^n \to X \cup cY$ with $\xi^k = Y \times I$, the trivial 1-dimensional bundle. Then f can be deformed to a map f_1 transverse to $Y \times ½$ and $V = f_1^{-1}(Y \times ½)$ splits M, $M - V = W \cup_\partial W'$, and after further deforming we get a map $\tilde{f}: (W,V) \to (X,Y)$. We set $\lambda\{M, f\} = \{W, \tilde{f}\}$, and it is not hard to see that λ is inverse to ϕ. Summarizing we have,

LEMMA 1.22. *If* (H)*-manifolds satisfy transversality then* $T_*^H(\ )$ *is a (generalized) homology theory.*

Recall that a spectrum $E = \{E_k, f_k\}$ consists of a sequence of based spaces E_k and (structure) maps $f_k : S^1 \wedge E_k \to E_{k+1}$. Associated to a spectrum there are (generalized) homology and cohomology theories ([4])

1.23
$$E_i(X) = \lim_{\substack{\to \\ k}} [S^{k+i}, X^+ \wedge E_k]^{*)},$$
$$E^i(X) = \lim_{\substack{\to \\ k}} [S^k \wedge X^+, E_{k+i}],$$

where $[-,-]$ denotes based homotopy classes and $X^+ = X \dot{\cup} (*)$.

Essentially every generalized homology theory (defined on finite CW complexes) is represented by a spectrum as in 1.23. This follows from Brown's representation theorem upon using Spanier-Whitehead duality (see e.g. [3]) to convert homology into cohomology. Thus under the assumptions of 1.22 we have a spectrum $MH = \{MH_k, f_k\}$ such that

$$T_n^H = \lim_{k \to \infty} \pi_{n+k}(MH_k) = \pi_n^S(MH)$$

giving an abstract homotopy interpretation of the cobordism groups T_n^H.

The spectrum MH above is the Thom spectrum for the universal (H)-bundles: Removing the zero section from γ_H^k we have a spherical fibration and we take MH_k to be its Thom space,

---

*) It is customary to denote $E_i(X)$ also by the symbol $\pi_i^S(X^+ \wedge E)$ where the s means stable homotopy group.

$$MH_k = \text{Mapping cone } (\gamma_H^k - BH_k \to BH_k)$$

THEOREM 1.24 (Thom). *Suppose the* (H)-*category satisfies transversality (of large codimension), then* $T_n^H(\ )$ *is represented by* MH,

$$T_n^H(X) = \lim_{k\to\infty} [S^{n+k}, X^+ \wedge MH_k] = \pi_n^S(X^+ \wedge MH)$$

Proof (Sketch). By assumption every homotopy class $[\gamma]$ of maps $\gamma: S^{k+n} \to X^+ \wedge MH_k$ (k large) can be represented by a map transverse to $X \times BH_k$. We get a diagram

1.25
$$\begin{array}{ccc} S^{k+n} & \xrightarrow{\gamma} & X^+ \wedge MH_k \\ \cup\!\uparrow & & \uparrow\!\cup \\ V^n & \xrightarrow{(f,\nu^k)} & X \times BH_k \end{array}$$

where ν^k is the (H)-normal bundle of V^n in S^{k+n} and $V^n = \gamma^{-1}(X \times BH_k)$. The pair (V^n, f) then represents the element of $T_n^H(X)$ associated to $[\gamma]$. Conversely, if (V^n, f) is a singular n-manifold in X then the associated homotopy class is represented by the composite

$$S^{k+n} \xrightarrow{c} M(\nu^k) \xrightarrow{\text{diag}} V^+ \wedge M(\nu^k) \xrightarrow{f \wedge M(\nu)} X^+ \wedge MH_k$$

where $V^n \subset S^{n+k}$ is an embedding with (H)-normal bundle ν^k, c the Pontrjagin-Thom collapse map onto a tubular neighborhood and $\nu: \nu^k \to \gamma_H^k$ the classifying map for ν^k.

Examples which satisfy transversality are PL-manifolds by the result of Williamson [146], and in dimension $n \neq 4$ topological manifolds by a result of Kirby-Siebenmann [65], as well as smooth manifolds from Thom's original results [137].

In the unoriented cases it is customary to denote the corresponding cobordism theories by $\mathfrak{N}_*^{PL}$, $\mathfrak{N}_*^{TOP}$ and $\mathfrak{N}_*$ and in the oriented cases by Ω_*^{PL}, Ω_*^{TOP} and Ω_*. From 1.24 we have

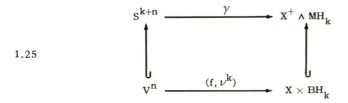

$$\Omega_*(X) = \pi_*^S(X^+ \wedge MSO)$$

1.26 $$\Omega_*^{PL}(X) = \pi_*^S(X^+ \wedge MSPL)$$

$$\Omega_*^{TOP}(X) = \pi_*^S(X^+ \wedge MSTOP), \qquad * \neq 4,$$

and similarly in the unoriented cases with MSO, MSPL and MSTOP replaced by MO, MPL and MTOP.

As a final example we have the almost complex manifolds. These are smooth manifolds M whose stable normal bundle is given a complex structure, that is, a homotopy class of liftings $\nu: M^n \to BU_k$ of the normal bundle $\nu: M^n \to BO_{2k}$ for k sufficiently large. The cobordism theory of almost complex manifolds is denoted Ω_*^U and again we have $\Omega_*^U(X) = \pi_*^S(X^+ \wedge MU)$.

Let K(A) denote the Eilenberg-MacLane spectrum for the Abelian group A with i'th space equal to K(A, i). Then, in the classic smooth categories we have ([137]).

THEOREM 1.27 (Thom). (a) *The spectrum* MO *has the homotopy type of a wedge of Eilenberg-MacLane spectra,* $MO \simeq \bigvee_{i=0}^{\infty} \Sigma^{n_i} K(Z/2)$.

(b) $\mathfrak{N}_*(pt) = P\{x_2, x_4, x_5, \cdots\}$ *with one generator in each dimension* $\neq 2^i - 1$ *and*

$$\mathfrak{N}_*(X) \cong H_*(X, \mathfrak{N}_*(pt)).$$

More generally we have the theorem of Browder, Liulevicius and Peterson [29]. Suppose we are given a notion of H_n bundles, weaker than O_n bundles (e.g. $H_n = TOP_n, PL_n$ or G_n). Then we have

THEOREM 1.28. i) *There is an algebra* C(H) *and*

$$\pi_*(MH) \cong \mathfrak{N}_*(pt) \otimes C(H).$$

ii) *If* H-*theory also satisfies transversality then*

$$\mathfrak{N}_*^H(X) \cong H_*(X, \mathfrak{N}_*^H(pt)) = H_*(X; \pi_*(MH)).$$

D. *Oriented cobordism theories and localization*

In the oriented case it is convenient to localize ([136]). We suppose X is an infinite loop space, i.e. we can write $X = \Omega^n Y_n$ for any n with Y_n $n-1$ connected, and the Y_n form a spectrum $\mathcal{Y}(X)$.

Let Λ be an abelian group. We define the Moore space $M(\Lambda)$ as the simply connected CW complex (unique up to homotopy type) having

$$H_2(M(\Lambda); \mathbf{Z}) = \Lambda$$

$$H_i(M(\Lambda); \mathbf{Z}) = 0 \qquad i > 2 .$$

Then we have

LEMMA 1.29. *There is an exact sequence*

$$0 \to \pi_*(X) \otimes \Lambda \to \pi^S_{*+2}(\mathcal{Y}(X) \wedge M(\Lambda)) \to \mathrm{Tor}(\pi_{*-1}(X), \Lambda) \to 0 .$$

Proof. Suppose A, B are free abelian groups and

$$0 \to A \overset{f}{\to} B \to \Lambda \to 0 \text{ *)}$$

a resolution of Λ then $M(\Lambda)$ is the cofiber

$$\bigvee_{\text{gen } A} S^2 \overset{\tilde{f}}{\to} \bigvee_{\text{gen } B} S^2 \to M(\Lambda)$$

when $\tilde{f}_* = f$. This gives a cofibering

$$\mathcal{Y}(X) \wedge \bigvee S^2 \to \mathcal{Y}(X) \wedge \bigvee S^2 \to \mathcal{Y}(X) \wedge M(\Lambda)$$

and 1.29 follows by applying the homology theory π^S_*.

*) As a slightly non-trivial example, for $\Lambda = \mathbf{Z}[1/p]$ the sequence takes the form

$$0 \to \bigoplus_{i=0}^{\infty} \mathbf{Z} \overset{f}{\to} \bigoplus_{i=0}^{\infty} \mathbf{Z} \to \mathbf{Z}[1/p] \to 0$$

where $f(z_0, z_1, \cdots) = (-z_0, pz_0 - z_1, pz_1 - z_2, \cdots)$.

In particular, if $\mathbf{Z}_{(p)}$ denotes the integers localized at p and $\mathbf{Z}[1/p]$ the integers localized away from p, i.e. $\mathbf{Z}_{(p)} = \{a/b \mid (b,p) = 1\}$, $\mathbf{Z}[1/p] = \{a/b \mid b = p^i \text{ some } i \geq 0\}$ then

$$\pi^s_{*+2}(\mathcal{Y}(X) \wedge M(\mathbf{Z}_{(p)})) = \pi_*(X) \otimes \mathbf{Z}_{(p)}$$

$$\pi^s_{*+2}(\mathcal{Y}(X) \wedge M(\mathbf{Z}[1/p])) = \pi_*(X) \otimes \mathbf{Z}[1/p]$$

and more generally

$$\pi^s_{*+2}(\mathcal{Y}(X) \wedge Z \wedge M(\Lambda)) = \pi^s_*(\mathcal{Y}(X) \wedge Z) \otimes \Lambda .$$

DEFINITION 1.30.*) (a) For X an infinite loop space we define $X[p]$, X localized at p as $\lim_{n \to \infty} \Omega^{n+2}(Y_n \wedge M(\mathbf{Z}_{(p)}))$ and $X[1/p]$, X localized away from p as

$$X[1/p] = \lim_{n \to \infty} \Omega^{n+2}(Y_n \wedge M(\mathbf{Z}[1/p])) .$$

(b) For a connected spectrum $\mathcal{Y}$ set $\mathcal{Y}[p]_i = \mathcal{Y}_{i-2} \wedge M(\mathbf{Z}_{(p)})$ and $\mathcal{Y}[1/p]_i = \mathcal{Y}_{i-2} \wedge M(\mathbf{Z}[1/p])$.

REMARK 1.31. The inclusions $\mathbf{Z} \to \Lambda$ (Λ as above) give maps $S^2 \to M(\Lambda)$ inducing $X \to X[\Lambda]$, $\mathcal{Y} \to \mathcal{Y}[\Lambda]$. They have the obvious effect in homotopy and homology of including $\pi_*(X[\Lambda]) \subset \pi_*(X) \otimes \Lambda$, etc.

By induction over a Postnikov system for X, $H_*(X[p]; \mathbf{Z}) \cong H_*(X, \mathbf{Z}) \otimes \mathbf{Z}_{(p)}$ and $H_*(X[1/p]; \mathbf{Z}) \cong H_*(X; \mathbf{Z}[1/p])$, the isomorphisms being induced from the inclusions in 1.31.

COROLLARY 1.32. *Let* $f: X \to Y$ *be a map of simply connected infinite loop spaces satisfying*

$$f_*: H_*(X; \Lambda) \xrightarrow{\cong} H_*(Y; \Lambda)$$

*) Alternately one can define $\mathcal{Y}[p]_i = \mathcal{Y}_i[p]$ where $\mathcal{Y}_i[p]$ is the space $\mathcal{Y}_i$ localized in the sense of [136].

for some Λ *above (and having the homotopy types of* CW *complexes) then* $f[\Lambda]: X[\Lambda] \to Y[\Lambda]$ *is a homotopy equivalence.*

EXAMPLE 1.33. The usual map $BSpin \to BO$ is a $Z[\frac{1}{2}]$ equivalence. Thus $MSpin \to MSO$ is a $Z[\frac{1}{2}]$ equivalence as well. Also $BSp \to BO$ is a $Z[\frac{1}{2}]$ equivalence where $BSp = \lim BSp_n$ and Sp_n is the symplectic group. On the other hand $BSp \times Z \simeq \Omega^4 BO$ by Bott periodicity, hence

$$\Omega^4(BO[\tfrac{1}{2}] \times Z[\tfrac{1}{2}]) \xrightarrow{\simeq} BO[\tfrac{1}{2}] \times Z[\tfrac{1}{2}]$$

and $BO[\frac{1}{2}] \times Z[\frac{1}{2}]$ is 4 fold periodic.

EXAMPLE 1.34. $X[Q]$ is X localized at Q (X an infinite loop space). It always has the homotopy type of a product of Eilenberg-MacLane spaces $K(Q;n)$, one for each primitive in $H_n(X,Q)$ provided X has the homotopy type of a locally finite CW complex. Similarly $\mathcal{Y}(X) \wedge M(Q)$ splits as a wedge of $K(Q)$ spectra (see e.g. [91, Appendix]).

Finally, from [95], [137], [142] we quote

THEOREM 1.35. (a) *The spectrum* $MSO[2]$ *has the homotopy type of a wedge of Eilenberg-MacLane spectra*

$$MSO[2] \simeq \bigvee \Sigma^{r_i} K(Z/2) \vee \bigvee \Sigma^{4s_j} K(Z_{(2)}) .$$

(b) *The torsion subgroup of* $\Omega_*(pt)$ *is a vector space over* $Z/2$ *and*

$$\Omega_*(X) \otimes Z_{(2)} \cong H_*(X; \Omega_*(pt) \otimes Z_{(2)}) .$$

(c) $\Omega_*(pt)/Tor = P\{x_4, x_8, \cdots\}$ *is a polynomial ring with one generator in each dimension* $4i$.

(d) $\Omega_*(pt) \otimes Q = P\{\{CP^2\}, \{CP^4\}, \cdots, \{CP^{2n}\}, \cdots\}$.

E. *Connections between cobordism and characteristic classes*

For notational convenience we set

$$F^*(X; A) = H^*(X; Z)/\mathrm{Tor} \otimes A$$
$$F_*(X; A) = H_*(X; Z)/\mathrm{Tor} \otimes A$$

where A is a subring of $\mathbf{Q}$. We suppress A when $A = \mathbf{Z}$.

The universal coefficient theorem identifies $F^*(X)$ with $\mathrm{Hom}(F_*(X), Z)$. Both F^* and F_* are natural and $F_*(\ ) \otimes A = F_*(\ ; A)$, $F^*(\ ) \otimes A = F^*(\ ; A)$. Also $F^*(X \times Y) = F^*(X) \otimes F^*(Y)$ and similarly for F_*. In particular

$$F^*(BO) = P\{p_4, p_8, \cdots, p_{4i}, \cdots\},$$

and the total Pontrjagin class

$$p = 1 + p_4 + p_8 + \cdots$$

is multiplicative,

$$\psi(p) = p \otimes p$$

where $\psi : F^*(BO) \to F^*(BO) \otimes F^*(BO)$ is the coproduct induced from Whitney sum. Equivalently, for bundles ξ, η we have $p(\xi \oplus \eta) = p(\xi) \cdot p(\eta)$ in $F^*(\text{base})$. (The total Pontrjagin class is not multiplicative in $H^*(\ ; Z)$ but the deviation is $Z/2$-torsion.)

More generally a (graded) class

$$\mathfrak{A} = 1 + A_4 + A_8 + \cdots + A_{4i} + \cdots$$

$A_{4i} \in F^*(BO)$ or $F^*(BO) \otimes \mathbf{Q}$ is said to be *multiplicative* or a *genus* if $\psi(\mathfrak{A}) = \mathfrak{A} \otimes \mathfrak{A}$.

Every genus is associated to its *characteristic* formal power series

1.36 $$S(\mathfrak{A}) = 1 + \sum \alpha_i x^i \qquad \alpha_i \in \mathbf{Q}$$

and conversely, any formal power series beginning with a 1 is associated to a genus ([54, Chapter 1]). Indeed if $\mathcal{A}$ is a genus then $S(\mathcal{A})$ is obtained by setting $p_{4i} = 0$ for $i > 1$ and $p_4 = x$ in $\mathcal{A}$. Conversely for the power series $f(x)$ set

$$F(\sigma_1, \sigma_2, \cdots) = \prod_{i=1}^{\infty} f(x_i)$$

where the σ_i are the elementary symmetric polynomials in the x_i (the leading term of $f(x)$ must be 1 for this to make sense). The genus associated to $f(x)$ is then

$$F(p_4, p_8, \cdots, p_{4n}, \cdots) \in F^*(BO) \otimes \mathbf{Q} .$$

$S(\mathcal{A})$ is not the only formal power series associated to $\mathcal{A}$. If we write $A_{4i} = \pi_{4i} p_{4i} + D$ where $\pi_{4i} \in \mathbf{Q}$ and D is decomposable, then we have the *primitive* series

1.37 $$P(\mathcal{A}) = \sum (-1)^i \pi_{4i} x^i \qquad (\pi_0 = 1)$$

and 1.36, 1.37 are related by the formulae

$$P(\mathcal{A}) = 1 - x \frac{d}{dx} \log S(\mathcal{A}) = 1 - x(S(\mathcal{A})'/S(\mathcal{A}))$$

$$S(\mathcal{A}) = \exp\left(\int \frac{1-P(\mathcal{A})}{x}\, dx\right)$$

all of these operations being carried out formally.

The set of multiplicative characteristic classes forms a group under multiplication

$$\mathcal{A} \cdot \mathcal{B} = \sum_k \left(\sum_{i+j=k} A_{4i} B_{4j} \right)$$

since formal power series with constant term 1 are formally invertible. Also, given $\mathcal{A}$ we can define a homomorphism again denoted $\mathcal{A}$,

$$\mathfrak{A} : \Omega_*(pt) \to Q$$

by $\mathfrak{A}\{M\} = <\mathfrak{A}^{-1}(\nu(M)), [M]>$. Since $\nu(M \times N) = \nu(M) \oplus \nu(N)$ we have $\mathfrak{A}\{M \times N\} = \mathfrak{A}\{M\} \cdot \mathfrak{A}\{N\}$ so $\mathfrak{A}$ is actually a ring homomorphism. Conversely,

THEOREM 1.38. *Given a ring homomorphism* $\mathfrak{A} : \Omega_*(pt) \to Q$ *there exists one and only one genus* $\mathfrak{B}$ *inducing it.*

Proof. $\mathfrak{A}$ extends uniquely to $\Omega_*(pt) \otimes Q \to Q$ and it suffices to check on a set of generators, which from 1.35(d) can be taken to be the CP^{2n}. Now $\mathfrak{B}^{-1}(\nu(CP^{2n})) = \mathfrak{B}(\tau(CP^{2n})) = \mathfrak{B}(\tau(CP^{2n}) + \varepsilon) = \mathfrak{B}((2n+1)H) = \mathfrak{B}(H)^{2n+1}$ where H is the canonical line bundle over CP^{2n}. Now $\mathfrak{B}(H) = \Sigma \beta_i d^{2i}$ where $S(\mathfrak{B}) = \Sigma \beta_i x^i$, and we must inductively solve for the β_i so that the coefficient of x^n in $S(\mathfrak{B})^{2n+1}$ is $\mathfrak{A}\{CP^{2n}\}$. Since this is uniquely possible 1.38 follows.

COROLLARY 1.39 (Hirzebruch index theorem). *Let* $\mathfrak{L}$ *be the genus associated to the characteristic series* $S(\mathfrak{L}) = \sqrt{x}/\tanh \sqrt{x}$.

Then, if M^{4n} *is an oriented differentiable manifold the index of* M *is equal to* $\mathfrak{L}\{M\}$.

Proof (sketch). Recall that index $I(M)$ is the signature of the cup product pairing on $H^{2n}(M, Q) \otimes H^{2n}(M, Q)$, $<a, b> = <a \cup b, [M]>$. It is first checked that $I(M)$ is a cobordism invariant and $I(M \times N) = I(M) \cdot I(N)$. Then, since $I(CP^{2n}) = 1$, the proof of 1.38 applies and 1.39 becomes equivalent to the assertion that the coefficient of x^n in $(\sqrt{x}/\tanh \sqrt{x})^{2n+1}$ is equal to 1. For a proof of this see [54].

Here is a second, slightly more delicate connection between characteristic classes and cobordism.

The generalized Hurewicz map

$$h : T_n^H(pt) \to H_n(BH)$$

is defined by $h\{M\} = \nu_*[M] \in H_*(BH; \Gamma)$ where Γ is Z for oriented theories and $Z/2$ for unoriented theories, and $\nu : M \to BH$ is the (H)-normal bundle. If the theory satisfies transversality, then $\{M, \nu\}$ is equivalent to a homotopy class $f: S^{n+k} \to MH_k$ $k >> n$, and $h(\{M, \nu\}) = \underline{h}(f) \cap U_k = f_*[S^{n+k}] \cap U_k$ where $\underline{h}$ is the usual Hurewicz map $\pi_*(X) \to H_*(X)$, and $U_k \in H^k(MH_k; \Gamma)$ the ordinary Thom class.

EXAMPLE 1.40. Let $\varepsilon : \Omega_*(pt) \to Z$ be the augmentation $\varepsilon(\{M\}) = 0$ if $\dim M > 0$ and $\varepsilon(\{M\}) = \# M$ if $\dim M = 0$. Then ε is a ring map, corresponding to the Thom class

$$U : MSO \to K(Z)$$

and the associated Hurewicz map of homology theories

$$h : \Omega_*(X) \to H_*(X; Z) ,$$

$$h(\{M, f\}) = f_*[M]$$

induces an isomorphism

$$\Omega_*(X) \otimes_{\Omega_*(pt)} Z_{(2)} \cong H_*(X; Z_{(2)}) .$$

EXAMPLE 1.41. The cobordism ring of almost complex manifolds $\Omega_*^U(pt)$ is torsion free and

$$\Omega_*^U(pt) \otimes Q = P\{\{CP^1\}, \cdots, \{CP^n\}, \cdots\}$$

(Milnor [95]).

From 1.16, $H_*(BU; Z) = P\{b_1, \cdots, b_n, \cdots\}$ and $\tau_*[CP^n]$ is the coefficient of x^n in $(1 + \Sigma b_i x^i)^{n+1}$. Also if $\chi : H_*(BU, Z) \to H_*(BU; Z)$ is the canonical (anti) automorphism induced from the map sending γ_U to $-\gamma_U$ then

$$h\{CP^n\} = \nu_*([CP^n]) = \left(1 + \sum \chi(b_i) x^i\right)_n^{n+1} .$$

Note that these remarks together with the preceding description of $\Omega_*(pt)$ give

THEOREM 1.42. *In* $\Omega_*(\mathrm{pt})$ *or* $\Omega_*^U(\mathrm{pt})$, M *and* M′ *determine the same element if and only if* $\nu_*[M] = \nu'_*[M']$.

REMARK 1.43. In the case of oriented PL manifolds or topological manifolds F. Peterson [110] has shown that there exist non-cobordant manifolds M^n with equal images $\nu_*[M^n]$ in $H_*(BSPL; Z)$ or $H_*(BSTOP; Z)$ so 1.42 does not extend to these cases. However, it is still of interest to characterize those classes in $H_*(BSPL; Z)$, $H_*(BSTOP; Z)$ as well as in $H_*(BSO; Z)$, $H_*(BU; Z)$ etc., which correspond to $\nu_*([M])$ for some M. In particular, in the latter two cases this characterization has taken the form of integrality theorems for certain complicated polynomials with rational coefficients in the Pontrjagin classes or Chern classes. This follows since the mapping $\phi : H^*(BSO; Q) \to \mathrm{Hom}(H_*(BSO; Z), Q)$ is an isomorphism, and ignoring torsion classes $\Omega_*(\mathrm{pt})/\mathrm{Tor} \subset H_*(BSO; Z)/\mathrm{Tor}$ is a sublattice which is completely determined by its dual lattice,

$$L^* = \{\phi \in \mathrm{Hom}(H_*(BSO; Z); Q) \mid \phi(\ell) \in Z \text{ for all } \ell \in L\}.$$

The elements in L^* are evidently rational polynomials in the Pontrjagin classes, and similarly for BU.

Thom in [138] defined Pontrjagin classes for PL-manifolds and $H^*(BSPL; Q) = P\{p_4, p_8, \cdots\}$, which we review in Chapter 4. From the work of Kirby-Siebenmann [65], $H^*(BSPL; Q) \cong H^*(BSTOP; Q)$ and we can ask for similar integrality theorems for PL and topological manifolds. We will return to these questions in Chapter 11.

CHAPTER 2

THE SURGERY CLASSIFICATION OF MANIFOLDS

In this chapter we review the Browder-Novikov theory and Sullivan's globalization of it which reduces many questions in the classification of manifolds to questions in homotopy theory.

The idea is to start the attempt at classification with a homotopy type which potentially could contain a manifold, find an obstruction theory to its actually being the homotopy type of a manifold and give an effective procedure for evaluating the obstructions. If one then knows that the homotopy type in question is represented by a manifold, relative versions of the theory serve to count the number of distinct manifolds in the given homotopy type. We limit ourselves to simply connected homotopy types. The reader is referred to [27] for a more detailed account.

A. *Poincaré duality spaces and the Spivak normal bundle*

DEFINITION 2.1. A simply connected finite CW complex X is said to be a Poincaré duality space (P.D. space) of (formal) dimension n if there is a class $[X] \in H_n(X; Z)$ so that

$$\cap[X] : H^i(X; Z) \to H_{n-i}(X; Z)$$

$(\cap[X] : \alpha \to \alpha \cap [X])$ is an isomorphism for all i.

Simply connected closed manifolds satisfy the conditions of 2.1, so a manifold determines a unique homotopy type of P.D. spaces.

A finite CW complex contains a finite simplicial complex in its homotopy type, and any such simplicial complex of geometric dimension m embeds in R^ℓ for ℓ large ($\ell \geq 2m+1$ suffices). We can then take a regular (or tubular) neighborhood of X by setting $N(X) = \text{Star}(X)$ in the second barycentric subdivision of the original triangulation of R^ℓ. From [116], X is a deformation retract of N(X) and $\partial N(X)$ is an $(\ell-1)$ dimensional triangulated manifold (that is, the star of every simplex is a PL-disc).

If X were a differentiable manifold and the embedding $X \subset R^{\ell}$ were differentiable then $N(X)$ would be identified with the normal disc bundle to X in R^{ℓ} and $\partial N(X)$ would be the normal sphere bundle.

This fact admits a basic generalization to the category of P.D. spaces ([127]).

THEOREM 2.2 (Spivak). *Let* X *be a simply connected P.D. space of (formal) dimension* n *and suppose* $X \subset R^{n+k}$ *with regular neighborhood* $N(X)$. *Then the homotopy fiber of the composite*

$$\partial N(X) \longrightarrow N(X) \xrightarrow[\simeq]{r} X$$

has the homotopy type of a $(k-1)$*-sphere.*

Thus, using Stasheff's classification theorem 1.3, corresponding to the fibering in 2.2 there is a classifying map

$$\nu^k : X \to BG_k \,.$$

Further analysis shows that the stable class ν of $\nu^k (k \to \infty)$ is independent of the choices of embedding and regular neighborhood. The resulting stable homotopy sphere bundle (also called spherical fibration) is the *Spivak normal bundle* to X.

As remarked above, in case X is already a smooth or piecewise linear manifold then $N(X)$ can be identified with the total space of the normal disc bundle to X and for k sufficiently large $\partial N(X) \to N(X) \to X$ is just the projection of the normal sphere bundle. In particular, in these cases ν is classified by a map into BO or BPL. Thus a necessary condition that a P.D. space X have the homotopy type of a smooth or PL manifold is that ν admits a reduction to BO or BPL.

The Thom complex $M(\xi)$ of a spherical fibration $\xi : S^{\ell-1} \to E \xrightarrow{\pi} X$, is the mapping cone of the projection π. There is a Thom class $U \in H^{\ell}(M(\xi); Z)$ and the cup product map

$$\Phi : H^i(X; Z) \to \tilde{H}^{\ell+i}(M(\xi); Z) ,$$

$\Phi(a) = a \cup U$ is an isomorphism (the Thom isomorphism). In the case of the Spivak normal bundle

$$M(\nu) = N(X)/\partial N(X) \simeq N(X) \cup c\,\partial N(X)$$

and $M(\nu \oplus \varepsilon^\ell) = \Sigma^\ell M(\nu)$, so the stable Spivak normal bundle is associated to the spectrum $\{M(\nu), \Sigma M(\nu), \Sigma^2 M(\nu), \cdots\}$.

If we collapse the complement of $N(X)$ in R^{m+k} we get a degree 1 map (the Pontrjagin-Thom map)

$$c : S^{n+k} \to M(\nu) .$$

Indeed, $c_*[S^{n+k}] \cap U = [X]$ for a suitable choice of Thom class U. The cap product is that dual to the cup product inducing Φ above, and

$$\lambda : H_{k+i}(M(\nu); Z) \to H_i(N(X); Z) = H_i(X, Z)$$

defined by $\lambda(a) = a \cap U$ is an isomorphism. One can now conveniently characterize the Spivak normal bundle: The isomorphism class of a stable spherical fibration γ^k over X contains a Spivak normal bundle if and only if there exists a degree 1 map $S^{n+k} \to M(\gamma^k)$. Moreover, if (γ_i^k, c_i), $i = 1, 2$ are pairs of stable spherical fibrations and degree 1 maps $c_i : S^{n+k} \to M(\gamma_i^k)$ then there exists a fiber homotopy equivalence $t : \gamma_1^k \to \gamma_2^k$ such that $M(t) \circ c_1 \simeq c_2 : S^{n+k} \to M(\gamma_2^k)$, and t is unique up to homotopy.

Recall that two finite CW complexes X and Y are called ℓ-dual if there are (large) suspensions $\Sigma^r X$ and $\Sigma^s Y$ and an embedding $\Sigma^r X \subset S^{\ell+r+s+1}$ with complement $S^{\ell+r+s+1} - \Sigma^r X$ homotopy equivalent to $\Sigma^s Y$. We have ([8]),

LEMMA 2.4. *The Thom space* $M(\nu^k)$ *is* $(n+k)$*-dual to* X_+.

Proof. Let $X \subset \text{int } D^{n+k}$ with regular neighborhood $N(X)$. Then $D^{n+k}/D^{n+k} - \text{int } N(X)$ is the Thom space $M(\nu^k)$. Moreover, as D^{n+k} is contractible $D^{n+k}/D^{n+k} - \text{int } N(X) \simeq \Sigma(D^{n+k} - \text{int } N(X))$. Now consider the embedding

$$X \subset D^{n+k} \subset S^{n+k} = D^{n+k} \cup c\, S^{n+k-1}\ .$$

Taking $(+)$ to be the cone point $S^{n+k} - X_+$ deforms onto $D^{n+k} - \text{int } N(X)$. But $\Sigma(S^{n+k} - X_+) \simeq S^{n+k+1} - X_+$ and the result follows.

We can use 2.4 to show the existence of a 5-dimensional simply connected P.D. space which does not have the homotopy type of a smooth manifold (see [78] for generalizations). Let X^5 be the 4-cell complex

2.5 $$X^5 = S^2 \vee S^3 \cup_f e^5, \qquad f = [\iota_2, \iota_3] + \eta^2 \iota_2\ .$$

Here, $\pi_4(S^2 \vee S^3) = \mathbf{Z} \oplus \pi_4(S^2) \oplus \pi_4(S^3)$ where the generator of the free summand is the Whitehead product $[\iota_2, \iota_3]$ (the attaching map of the top cell in the Cartesian product $S^2 \times S^3$), $\pi_4(S^2) = \mathbf{Z}/2$ with generator η^2 $(\eta^2 = \eta \circ \sigma\eta)$ and $\pi_4(S^3) = \mathbf{Z}/2$ with generator $\sigma\eta$. The presence of $[\iota_2, \iota_3]$ in the attaching map of e^5 shows that $e^2 \cup e^3 = e^5$ in cohomology, so X^5 is a P.D. space.

On the other hand since the suspension of any Whitehead product is zero we have

$$\Sigma^n(X^5_+) = (S^{n+2} \cup_{\sigma^n(\eta^2)} e^{n+5}) \vee S^{n+3} \vee S^n$$

and $\sigma^n(\eta^2) \in \pi_{n+4}(S^{n+2}) = \mathbf{Z}/2$ is the non-trivial class. Thus the Thom space of the Spivak normal bundle is the dual of $\Sigma^n(X^5_+)$ and can be written

2.6 $$Y = (S^\ell \cup_g e^{\ell+3}) \vee S^{\ell+2} \vee S^{\ell+5}\ .$$

Here $g \in \pi_{\ell+2}(S^\ell) = \mathbf{Z}/2$ is again the non-trivial element, $g = \sigma^{\ell-2}(\eta^2)$.

We show that Y is *not* the homotopy type of the Thom space for any vector bundle. First, note that since $Sq^i \equiv 0$ in $H^*(X^5)$ for all $i > 0$ the total Wu class V (see footnote on p. 80) of X^5 is 1 and since $Sq(V) = W$ —the total

Stiefel-Whitney class of X^5 —we see that $W = 1$ and $W^{-1} = 1$ which is the total Stiefel-Whitney class of the Spivak normal bundle $\nu(X^5)$. Consequently $w_2(\nu(X^5)) = 0$ and if ξ is a vector bundle over X^5 with $M(\xi) = Y$ then ξ is classified by a map

$$h : X^5 \to \mathrm{BSpin}_\ell$$

Consider the induced map

$$M(h) : Y = M(\xi) \to M(\gamma^\ell_{\mathrm{Spin}})$$

where $\gamma^\ell_{\mathrm{Spin}}$ is the universal bundle over BSpin_ℓ. The secondary operation Φ based on the relation

$$Sq^2 Sq^2 + Sq^3 Sq^1 = 0$$

(cf. [1]) is defined on the Thom class U in $H^\ell(M\gamma_{\mathrm{Spin}}; \mathbf{Z}/2)$ and it has value zero since $H^{\ell+3}(M\gamma_{\mathrm{Spin}}; \mathbf{Z}/2) \simeq H^3(\mathrm{BSpin}; \mathbf{Z}/2) = 0$. But since Φ detects $\sigma^{\ell-2}(\eta^2)$, Φ is non-zero in Y and this gives a contradiction.

REMARK 2.7. The fiber PL/O in the fibration

$$\mathrm{PL/O} \to \mathrm{BSO} \to \mathrm{BSPL}$$

is 6-connected by the result of Cerf [39]. Hence $[X^5, \mathrm{BSPL}] \cong [X^5, \mathrm{BSO}]$ and the above arguments show that X^5 is not homotopy equivalent to a PL-manifold either. Also, using deeper results of Kirby-Siebenmann it follows that X^5 does not even have the homotopy type of a topological manifold.

REMARK 2.8. We can classify the remaining Poincaré duality spaces of dimension 5 with 4-skeleton $S^2 \vee S^3$ as follows:

a) $$(S^2 \vee S^3) \cup_{[\iota_2,\iota_3]} e^5 = S^2 \times S^3 ,$$

b) $$(S^2 \vee S^3) \cup_{[\iota_2,\iota_3]+(\sigma\eta)\iota_3} e^5 = S(H \oplus \varepsilon^2)$$

b′) $$(S^2 \vee S^3) \cup_{[\iota_2,\iota_3]+\eta^2\iota_2+(\sigma\eta)\iota_3} e^5 \simeq S(H \oplus \varepsilon^2) .$$

Here H is the Hopf bundle over S^2, ε^2 the 2-dimensional trivial bundle, and $S(H \oplus \varepsilon^2)$ is the total space of the associated sphere bundle. The homotopy equivalence in b′) comes from the homotopy equivalence

$$\iota_2 \vee (\eta\iota_2 + \iota_3) : S^2 \vee S^3 \to S^2 \vee S^3$$

which takes the attaching map $[\iota_2, \iota_3] + (\sigma\eta)\iota_3$ to $[\iota_2, \iota_3] + \eta^2\iota_2 + (\sigma\eta)\iota_3$.

We have seen above that a 1'st obstruction to a P.D. space containing a smooth or PL manifold in its homotopy type is the existence of a reduction of its Spivak normal bundle to BPL or BO,

$$\begin{array}{ccc} & & BO \\ & \overset{\xi}{\nearrow} & \big\downarrow \pi \\ X & \xrightarrow{\ \nu\ } & BG \end{array} \qquad\qquad \begin{array}{ccc} & & BPL \\ & \overset{\xi}{\nearrow} & \big\downarrow \pi' \\ X & \xrightarrow{\ \nu\ } & BG \end{array}$$

Such a reduction is often called a *normal invariant* and two such are equivalent if the liftings are homotopic (by a homotopy constant when projected to BG).

REMARK 2.9. A basic result of Boardman-Vogt [19] shows that π, π' above are the homotopy types of principal fiberings and there are fibrations

$$BO \xrightarrow{\pi} BG \xrightarrow{\rho} B(G/O), \qquad BPL \xrightarrow{\pi} BG \xrightarrow{\rho'} B(G/PL).$$

Thus the obstruction to the existence of a smooth or PL normal invariant for a P.D. space X is an element in $[X, B(G/O)]$ or in $[X, B(G/PL)]$.

B. *The Browder-Novikov theorems and degree* 1 *normal maps*

To what extent does the existence of a normal invariant guarantee a manifold in a given homotopy type? This leads to a 2nd (and actually secondary) obstruction theory—the surgery obstructions.

Let ξ^k be a reduction of the stable Spivak normal bundle to BO or BPL i.e. ξ^k is a vector bundle or PL R^k-bundle with a proper fiber homotopy equivalence $\xi^k \to \nu^k$. Then we have an identification $M(\xi^k) \xrightarrow{\simeq} M(\nu^k)$ and the Pontrjagin-Thom map gives a degree 1 map

$$c: S^{n+k} \to M(\xi^k)\,.$$

We can deform c to a map c_1 transverse to $X \subset M(\xi^k)$ and then $M^n = c_1^{-1}(X)$ becomes a codimension k smooth or PL submanifold of S^{n+k} whose normal bundle is mapped to ξ^k by a bundle map (cf. Chapter 1). Thus we have the diagram

$$\begin{array}{ccc} \nu_{(M)} & \xrightarrow{\hat{f}} & \xi \\ \downarrow & & \downarrow \\ M^n & \xrightarrow{f} & X \end{array} \qquad (2.10)$$

($f = c_1|M$, $\hat{f} = c_1|$ tubular neighborhood) and f has degree 1, $f_*([M]) = [X]$.

DEFINITION 2.11. The diagram 2.10 is called a normal map or surgery problem over X, ξ. If $f_*([M]) = [X]$ then we say it is a degree 1 normal map.

Later in this section we shall also consider normal maps for manifolds or P.D. spaces with boundary. The definition is as in 2.11 with the extra assumption that $f|\partial M : \partial M \to \partial X$ is a homotopy equivalence (or sometimes a diffeomorphism or PL-homeomorphism).

The uniqueness of the Spivak normal bundle has the following consequence.

COROLLARY 2.12. *Let* X *be a P.D. space and* ξ *a vector bundle or* PL *bundle over* X. *Then there is a degree* 1 *normal map with range* X, ξ *if and only if* ξ *is fiber homotopy equivalent to the Spivak normal bundle.*

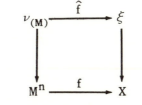

A reduction of the Spivak normal bundle to BO or BPL thus leads to a smooth or PL degree 1 normal map; but this correspondence is not well defined. For one thing, two reductions $\xi_i \to \nu$, $i = 1, 2$ are counted equal if there exists an isomorphism $\xi_2 \to \xi_1$ (linear or PL) taking one into the other. More seriously, in the process we deformed the degree 1 map $c : S^{n+k} \to M(\xi)$ to a transverse map and the resulting normal map is not independent of which transverse deformation of c we pick. However, a relative version of transversality shows that two homotopic transverse maps c_0, c_1 lead to cobordant normal maps. Precisely

DEFINITION 2.13. Two normal maps $f_i : M_i \to X$, $\hat{f}_i : \nu(M_i) \to \xi_i$, $i = 1, 2$, are called normally cobordant if there exists a W, $\partial W = M_1 \dot{\cup} M_2$ and a diagram (where $\hat{F}$ is a bundle map)

$$\begin{array}{ccc} \nu(W) & \xrightarrow{\hat{F}} & \xi \times I \\ \downarrow & & \downarrow \\ W & \xrightarrow{F} & X \times I \end{array}$$

so that $(F|M_1, \hat{F}|\nu(M_1)) = (f_1, \hat{f}_1)$ and $(F|M_2, \hat{F}|\nu(M_2)) = (f_2, b \circ \hat{f}_2)$ where $b : \xi_2 \to \xi_1$ is a bundle isomorphism.

Normal cobordism is an equivalence relation and the set of equivalence classes with range X is denoted $NM_O(X)$ or $NM_{PL}(X)$ in the smooth or PL case, respectively. Now, it is not hard to see that the normal invariants of X are in 1-1 correspondence with the elements of $NM(X)$.

EXAMPLE 2.14. Let $f : M \to X$ be a homotopy equivalence where M is a smooth or PL manifold. If $g : X \to M$ is a homotopy inverse to f set $\xi = g^*(\nu(M))$ so $f^*(\xi) \cong \nu(M)$. Thus f becomes part of a degree 1 normal map whose normal cobordism class is well defined.

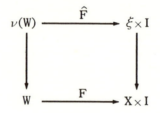

Given a normal cobordism class of normal maps over a simply connected P.D. space X we can ask if it has a representative $(f, \hat{f})$ with f a homotopy equivalence. The process of surgery ([27]) insures that we can make f highly connected. In fact for $n = \dim X \geq 5$ f can be assumed to be a homotopy equivalence for n odd and $(n/2)-1$ connected if n is even. The obstruction to completing surgery is an Arf invariant with values in $Z/2$ for $n \equiv 2(4)$ and $I(M)-I(X) \in 8Z$ if $n \equiv 0(4)$ where $I(M)$ denotes the index (= signature) of the cup product pairing $H^{n/2}(M;R) \otimes H^{n/2}(M;R) \to R$ and similarly for $I(X)$. This is summarized in

THEOREM 2.15 (Browder-Novikov). *Let X be an n-dimensional simply connected Poincaré duality space, $n \geq 5$ and suppose that its Spivak normal bundle ν admits a reduction to BO. Then for n odd X is homotopic to a smooth manifold and if n is even the obstruction is an element in $Z/2$ for $n \equiv 2(4)$ and in Z for $n \equiv 0(4)$. If $n \equiv 0(4)$ the obstruction vanishes if and only if for some reduction ξ of ν, $\langle \mathcal{L}^{-1}(\xi), [X] \rangle = I(X)$.*

Note that $\langle \mathcal{L}^{-1}(\xi), [X] \rangle = \langle \mathcal{L}^{-1}(\xi), f_*[M] \rangle = \langle \mathcal{L}^{-1}(\nu(M)), [M] \rangle = \langle \mathcal{L}(\tau M), [M] \rangle = I(M)$ by the Hirzebruch index theorem, (cf. 1.39).

REMARK 2.16. In Chapter 8.B. we review the construction of smooth manifolds $\overset{\circ}{M}{}_i^{4n}$ with boundary a homotopy sphere Σ^{4n-1} $(n>1)$ and index $8i$ $(i \neq 0)$ and of smooth manifolds $\overset{\circ}{M}{}^{4n+2}$ with boundary a homotopy $(4n+1)$-sphere and having "Kervaire invariant one". We call these manifolds the open Milnor manifolds and open Kervaire manifolds. Their normal bundles are trivial and hence induced from a degree 1 map onto a disc obtained as the map pinching the complement of a collar

$$f : (\overset{\circ}{M}{}_i^{4n+2\varepsilon}, \Sigma_i^{4n+2\varepsilon-1}) \to (D^{4n+2\varepsilon}, S^{4n+2\varepsilon-1}), \qquad \varepsilon = 0, 1$$

from the trivial bundle. Hence f is part of a smooth degree 1 normal map (with boundary) $(f, \hat{f})$. When $n=1$ the desingularized Kummer surface K^4([105], [126]) is simply connected, has index 16 and even cup pairing $H^2(K) \otimes H^2(K) \to Z$. Thus $w_2(K) = 0$ and K is again a Spin manifold, so $\overset{\circ}{K} = K - \overset{\circ}{D}{}^4$ is parallelizable. Connected sum along the boundary (see the remarks after 2.18) then give degree 1 normal maps over (D^4, S^3) with index $16i$ for all i. There is no closed differentiable (or PL) Spin manifold of dimension 4 and index 8 so this is the best we can do.

The underlying PL-manifold of $\Sigma_i^{4n+2\varepsilon-1}$ is PL-homeomorphic to $S^{4n+2\varepsilon-1}$ by the h-cobordism theorem ([116]). Thus we can cone it off to obtain a closed PL-manifold $M_i^{4n+2\varepsilon} = \mathring{M}_i^{4n+2\varepsilon} \cup c\,\Sigma^{4n+2\varepsilon-1}$ of index $8i$ or "Kervaire invariant 1". The PL-normal bundle to $M^{4n+2\varepsilon}$ is again induced by the pinching map

$$f : M_i^{4n+2\varepsilon} \to S^{4n+2\varepsilon}$$

this time from an (in general) non-trivial PL-bundle ξ over $S^{4n+2\varepsilon}$, so again f is part of a PL degree 1 normal map.

It is known that M_i^{4n} is differentiable if and only if i is divisible by $a_n 2^{2n-2}(2^{2n-1}-1)\,\mathrm{Num}(B_{2n}/4n)$ where B_{2n} is the Bernoulli number (cf. Chapter 11) and $a_n = 1$ for n even and $a_n = 2$ for n odd ([62]). Also, the results of [26] show that M^{4n-2} is not differentiable if n is not a power of 2.

The Browder-Novikov theorem 2.15 remains valid in the PL-category. Here, however, one can do better.

COROLLARY 2.17. *Let* X *be a simply connected P.D. space whose Spivak normal bundle admits a reduction to* BPL. *Then* X *is homotopic to a* PL *manifold if* $\dim X \geq 5$.

Proof. As before we have a degree 1 normal map $f: M^n \to X$, $\hat{f}: \nu(M) \to \xi$. Suppose $n \equiv 0(4)$ and that $I(M) - I(X) = 8i$. Then we take the connected sum of $(f, \hat{f})$ with the Milnor surgery problem $M^n_{-i} \to S^n$ of index $-8i$ (2.16) to get a surgery problem over $X \# S^n = X$ with vanishing surgery obstruction. If $n \equiv 2(4)$ we use the connected sum with the Kervaire surgery problem.

C. *The number of manifolds in a homotopy type*

Novikov gave a method for counting the number of possible manifolds in the homotopy type in terms of the number of distinct liftings of the normal bundle. Several people have, since, sharpened this result. We now consider this question of classifying the manifolds within a fixed homotopy type.

To begin we recall from 2.14 that associated to a homotopy equivalence $f: M^n \to X$ there is a unique normal cobordism class of degree 1 normal maps $(f, \hat{f})$. The next result is a relative version of 2.15.

THEOREM 2.18 (Novikov). *Suppose two homotopy equivalences* $f_i : M_i^n \to X$, $i = 1, 2$ *give normally cobordant normal maps. If* M_1^n *and* M_2^n *are smooth (and* $\pi_1(X) = 0$) *then* M_1^n *is diffeomorphic to the connected sum* $M_2^n \# \Sigma^n$ *of* M_2^n *with a homotopy sphere bounding an open Milnor or Kervaire manifold. In the* PL-*case* M_1^n *is* PL-*homeomorphic to* M_2^n. *In both cases we have assumed* $n \geq 5$.

First, recall the operation of connected sum along the boundary: Let $H^{n+1} = \{(x_1, \cdots, x_{n+1}) | \Sigma x_i^2 \leq 1, x_1 \geq 0\}$ and let $D^n \subset H^{n+1}$ be the n-disc of points with $x_1 = 0$. For manifolds W_1^{n+1}, W_2^{n+1} with boundary we choose embeddings $(H^{n+1}, D^n) \subset (W_i, \partial W_i)$, orientation preserving for $i = 1$ and orientation reversing for $i = 2$, and set

$$W_1^{n+1} \#_\partial W_2^{n+1} = (W_1 - H^{n+1}) \cup (W_2 - H^{n+1}) .$$

The diffeomorphism (or PL-homeomorphism) type of $W_1^{n+1} \#_\partial W_2^{n+1}$ is well defined, D^{n+1} acts as the identity and

$$\partial(W_1^{n+1} \#_\partial W_2^{n+1}) = \partial W_1 \# \partial W_2 .$$

Proof of 2.18. Let $F : W^{n+1} \to X \times I$ be the normal cobordism between the normal maps $f_i : M_i^n \to X$. We attempt to do surgery on the interior of W to obtain a homotopy equivalence $F' : W' \to X \times I$. This is always possible if $n+1$ is odd but if $n+1$ is even we may have an obstruction in $\mathbf{Z}$ or $\mathbf{Z}/2$ according to the parity of $(n+1)/2$. But if this happens we can add a suitable open Milnor manifold or Kervaire manifold $\mathring{M}^{n+1}$ to W along the boundary component M_2 of W to replace the original normal map $(F, \hat{F})$ with a normal map $(F', \hat{F}')$, $F' : W' \to X \times I$ on which interior surgery may be completed. Hence there is a homotopy equivalence $F'' : W'' \to X \times I$

where $\partial W'' = M_1 \dot{\cup} (M_2 \# \Sigma^n)$. Since $F'': W'' \to X \times I$ is a homotopy equivalence W'' is an h-cobordism and from the h-cobordism theorem ([116], [97]) we have $M_1 \cong M_2 \# \Sigma^n$ as required. The PL-case is similar.

Let X be a simply connected smooth or PL manifold, fixing a homotopy type. Following Sullivan we now make the definition

DEFINITION 2.19. A homotopy equivalence $f: M^n \to X$ where M is smooth or PL is called a homotopy smoothing or homotopy triangulation of X. Two such

$$f_1 : M_1 \to X, \quad f_2 : M_2 \to X$$

are equivalent if there is an h-cobordism W, $\partial W = M_1 \dot{\cup} M_2$ and a map $F: W \to X$ so that $F|M_i = f_i$. The set of equivalence classes is denoted $\mathcal{S}_O(X)$ and $\mathcal{S}_{PL}(X)$, respectively.

From 2.14 there are maps

$$\mathcal{S}_O(X) \xrightarrow{j} NM_O(X), \quad \mathcal{S}_{PL}(X) \xrightarrow{j} NM_{PL}(X)$$

and an element $\alpha \in NM(X)$ is in the image of j if and only if the surgery invariant associated to α is zero. This may be restated as an exact sequence of sets

2.20

$$\mathcal{S}_O(X) \xrightarrow{j} NM_O(X) \xrightarrow{s} P_n,$$

$$\mathcal{S}_{PL}(X) \xrightarrow{j} NM_{PL}(X) \xrightarrow{s} P_n, \qquad n = \dim X \geq 5$$

where P_* denotes the simply connected surgery groups, $P_n = Z$ for $n \equiv 0(4)$, $P_n = Z/2$ for $n \equiv 2(4)$ and $P_n = 0$ for n odd. (See Chapter 4 for the definition of the maps s.)

The obstruction groups P_n are the same in the smooth and PL case, but the obstruction maps are different. From 2.16 $s: NM_{PL}(X) \to P_n$ is onto but this is not true in the smooth case.

REMARK 2.21. The sequences 2.20 can be continued to the left. First, P_{n+1} acts on $\mathcal{S}(X)$ and j is an injection on orbits. Indeed given $\alpha \in P_{n+1}$ and a homotopy equivalence f one can find a degree 1 normal map $(F,\hat{F})$, $F: W^{n+1} \to X\times I$ such that $\partial W = M \dot{\cup} M'$, $F|M = f$ and $F|M' = f'$ is a homotopy equivalence, and such that the obstruction to doing surgery on the interior of W to get a homotopy equivalence is precisely α. (In fact, one can take the open Kervaire manifold or Milnor manifold $\overset{\circ}{M}{}_i^{n+1}$ representing α and set

$$W = (M\times I) \#_{\partial} \overset{\circ}{M}{}_i^{n+1}$$

where the connected sum takes place along $M\times(1)$.) One then defines $\alpha\cdot\{f\}=\{f'\}$. In the smooth case this takes $f: M^n \to X$ to $f' = f \# p: M^n \# \Sigma^n \to X \# S^n = X$ where p is the pinching map $p: \Sigma^n \to S^n$ and in the PL-case the operation is trivial.

To further continue 2.20 to the left one needs relative groups, so define $NM_O(V,\partial V)$, $NM_{PL}(V,\partial V)$ to be the equivalence classes of degree 1 normal maps, $\tilde{V} \to V$ which are diffeomorphisms, PL-homeomorphisms on the boundary. The normal cobordism relation has $\partial\tilde{W} = \tilde{V}_1 \cup \tilde{V}_2 \cup (\partial\tilde{V}_1\times I)$ with $\partial\tilde{V}_1 \times 1$ identified with $\partial\tilde{V}_2$. Similarly we define $\mathcal{S}_{PL}(V,\partial V)$ and $\mathcal{S}_O(V,\partial V)$. It is now easy to check the exactness of

$$\cdots \longrightarrow NM_O(X\times I,\partial) \xrightarrow{s} P_{n+1} \longrightarrow \mathcal{S}_O(X) \longrightarrow NM_O(X) \xrightarrow{s} P_n$$

$$\cdots \longrightarrow NM_{PL}(X\times I,\partial) \xrightarrow{s} P_{n+1} \longrightarrow \mathcal{S}_{PL}(X) \longrightarrow NM_{PL}(X) \xrightarrow{s} P_n$$

where X is an n-dimensional smooth or PL manifold, $n\geq 5$ (cf. [141]).

REMARK 2.22. Let $\mathrm{Aut}(X)$ denote the group of homotopy classes of homotopy equivalences of X. This group acts on $\mathcal{S}(X)$ and the orbit set $\mathcal{S}(X)/\mathrm{Aut}(X)$ corresponds to the set of diffeomorphism or PL-homeomorphism types contained in the homotopy type determined by X. The groups $\mathrm{Aut}(X)$ are hard to compute in general and very little is known about them except for the general result of Sullivan's: $\mathrm{Aut}(X)$ is an arithmetic group.

Let G/O and G/PL be the fibers in the fibrations

$$G/O \longrightarrow BO \xrightarrow{\pi} BG$$

$$G/PL \longrightarrow BPL \xrightarrow{\pi'} BG$$

Then we have the following fundamental result (Sullivan [134], Lashof-Rothenberg [70]) which ultimately makes $NM_{PL}(X)$ and to some extent $NM_O(X)$ computable.

THEOREM 2.23. (a) *If* X *is a smooth manifold then*

$$NM_O(X) = [X, G/O] .$$

(b) *If* X *is a* PL *manifold then* $NM_{PL}(X) = [X, G/PL]$.

Proof. A map $X \to G/PL$ is equivalent to specifying a PL-bundle λ over X together with a proper fiber homotopy equivalence

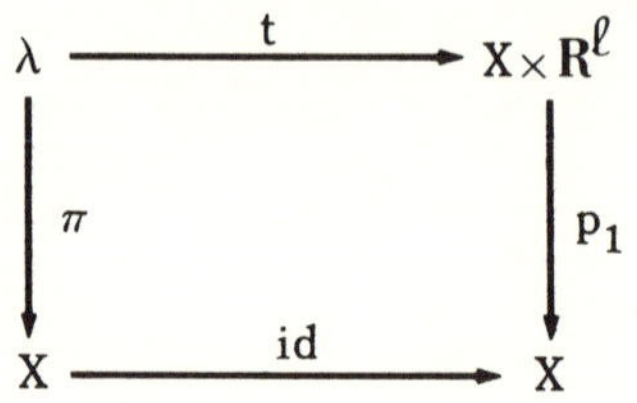

Now deform t to a map transverse to $X \times 0$. The inverse image of X is M and $\pi|M: M \to X$ has degree 1 by Poincaré duality. The normal bundle to M in λ is framed and the normal bundle to λ is $\pi^*(\nu) - \pi^*(\lambda)$. Hence the normal bundle to M is identified with $\pi^*(\nu(X) - \lambda)$ so we have a degree 1 normal map.

Conversely, given a degree 1 normal map as in 2.10 set $\lambda = \nu(X) - \xi$, then making the dimension of λ sufficiently big there is an embedding $f': M \hookrightarrow \lambda$ giving a commutative diagram

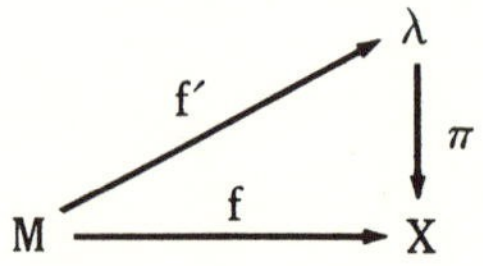

The normal bundle to λ is ξ and since $\nu(M) = f^*(\xi)$ it follows that the normal bundle to $f'(M)$ in λ is trivialized. (We assume the fiber dimension sufficiently large to insure a normal PL-bundle and so large that this bundle is stable.) Hence we obtain an extension of f', $M \times D^{\ell} \hookrightarrow \lambda$. The Pontrjagin-Thom construction gives a map

$$\mu : M(\lambda) \to \Sigma^{\ell}(M_+) .$$

Poincare duality and the fact that f has degree 1 implies

$$< i^* \mu^*(U^{\ell}), [S^{\ell}] > = 1$$

where $i : S^{\ell} \to M(\lambda)$ is the inclusion of a fiber. Let $p : \Sigma^{\ell}(M_+) \to S^{\ell}$ be the projection which collapses M to a point, then as $M(\lambda) = S(\lambda \oplus 1)/X$,

$$\pi \times p\mu : S(\lambda \oplus 1) \to X \times S^{\ell}$$

becomes a homotopy equivalence and thus specifies a map into G/PL.

Checking well definedness up to homotopy and that the two constructions are mutually inverse is direct. The argument is similar in the smooth case.

COROLLARY 2.24. $\pi_n(G/PL) = P_n$, *where* P_n *is as above.*

Proof. For $n \geq 5$ we have the exact sequence of sets

$$\mathcal{S}_{PL}(S^n) \longrightarrow \pi_n(G/PL) \xrightarrow{s} P_n .$$

But $\mathcal{S}_{PL}(S^n) = *$ by the generalized Poincaré conjecture, s is onto (2.16) and it is not hard to see that s is additive.

For $n \leq 5$ we use that PL/O is 6-connected ([39]) so that $G/O \to G/PL$ is a homotopy equivalence in the range under study. Now a direct check gives $\pi_4(G/O) = \mathbf{Z}$, $\pi_2(G/O) = \mathbf{Z}/2$ and $\pi_3(G/O) = \pi_1(G/O) = 0$.

REMARK 2.25. So far we have only considered smooth and PL manifolds. However, the work of Kirby-Siebenmann [65] insures that the results are also valid for topological manifolds. In particular there is an exact sequence

$$* \to \mathcal{S}_{TOP}(X) \to [X, G/TOP] \to P_n \to 0$$

for each topological manifold X of dimension at least 5.

In the fibration

$$TOP/PL \to G/PL \to G/TOP$$

the fiber TOP/PL is an Eilenberg-MacLane space $K(Z/2, 3)$ and the homotopy sequence

$$0 \longrightarrow \pi_4(G/PL) \longrightarrow \pi_4(G/TOP) \xrightarrow{\partial_*} \pi_3(TOP/PL) \longrightarrow 0$$

is non-split. Thus $\pi_n(G/PL) \cong \pi_n(G/TOP)$ and the natural map induces the isomorphism except in degree 4 where it induces multiplication by 2.

CHAPTER 3
THE SPACES SG AND BSG

From the previous chapters we have seen the importance of the spaces of homotopy equivalences and oriented (degree 1) homotopy equivalences of the n-sphere, G_{n+1} and SG_{n+1} respectively, in classifying manifolds. In this chapter we study the homotopy types and the mod. 2 cohomology of these spaces and their classifying spaces.

A. *The spaces of stable homotopy equivalences*

The equatorial inclusion $S^n \subset S^{n+1}$ induces inclusions $i : G_{n+1} \to G_{n+2}$, $i : SG_{n+1} \to SG_{n+2}$,

$$\begin{aligned} i(f)(t_0, \cdots, t_{n+1}) &= (t_0, \lambda^{-1} \cdot f(\lambda t_1, \cdots, \lambda t_{n+1})), & t_0 \neq \pm 1 \\ &= (t_0, 0, \cdots, 0), & t_0 = \pm 1 \end{aligned}$$

where $\lambda = 1/\sqrt{1-t_0^2}$. Clearly, $i(f) \circ i(g) = i(f \circ g)$ so i is a homomorphism of monoids and induces maps of classifying spaces

$$Bi : BG_{n+1} \to BG_{n+2}, \quad Bi : BSG_{n+1} \to BSG_{n+2} .$$

The universal S^{n+1}-bundle γ^{n+2} over BG_{n+2} pulls back to the Whitney sum (i.e. fiberwise join) of γ^{n+1} and the 1-dimensional trivial bundle. Hence if we set

$$BG = \lim_{\to} BG_n, \quad BSG = \lim_{\to} BSG_n$$

we see from Stasheff's classification theorem (cf. 1.3) that for X a finite complex the homotopy set $[X, BG]$ (respectively $[X, BSG]$) is in one to one correspondence with the stable equivalence classes of homotopy (respectively oriented homotopy) sphere bundles over X.

We now study the homotopy types of G and BG. Let $* \in S^n$ be a base point (usually we take $* = (1, 0, \cdots, 0)$) and consider the evaluation map

$$E : G_{n+1} \to S^n, \quad E(f) = f(*) .$$

We claim that E is a fibration.

A map $f: X \to Y$ satisfies the path lifting property (PLP) if for each space K and homotopy $h_t : K \times I \to Y$ whose initial term is lifted to $H_0 : K \to X$ $(f \circ H_0 = h_0)$ there exists a homotopy $H_t : K \times I \to X$ covering $h_t (f \circ H_t = h_t)$. (Also recall that each map $f: X \to Y$ can be converted into a fibration and thus has a homotopy (theoretical) fiber. If f satisfies PLP then $f^{-1}(*)$ is equal to the homotopy fiber.)

LEMMA 3.1. *The map* $E : G_{n+1} \to S^n$ *satisfies* PLP *and* $E^{-1}(*) = F_n$, *consisting of all those* $f \in G_{n+1}$ *with* $f(*) = *$.

(See e.g. [125] for a proof.)

The inclusion $i : G_{n+1} \to G_{n+2}$ takes G_{n+1} into F_{n+1}. Of more importance, however, is the inclusion $F_n \to G_{n+1} \to F_{n+1}$ which we analyze now.

Recall that if X is a space with base point $*$ then the loop space of X, ΩX, is the space of all base point preserving maps $(S^1, *) \to (X, *)$ equipped with the compact-open topology. The set of base point preserving continuous maps $f: S^1 \wedge Y \to X$ is in one to one correspondence with base point preserving continuous maps $g: Y \to \Omega X$,

3.2 $$\text{Maps}(S^1 \wedge Y, X) \approx \text{Maps}(Y, \Omega X) ,$$

where the loop in X with constant value $*$ serves as base point for ΩX. To f one associates the adjoint $g(y)(t) = f(t, y)$ and vice versa. The identification induces an identification of based homotopy sets

3.3 $$[S^1 \wedge Y, X] \approx [Y, \Omega X] .$$

Iterating 3.2 we have, $\text{Maps}(S^n \wedge Y, X) \approx \text{Maps}(Y, \Omega^n(X))$ where

$\Omega^n X = \Omega(\Omega^{n-1}X)$. In particular $\Omega^n X$ is the set of base point preserving maps $(S^n, *) \to (X, *)$ and $\pi_i(\Omega^n X, *) = \pi_{i+n}(X, *)$ for all $i \geq 0$. As a special case of this last relation, note that $\pi_0(\Omega^n X, *)$ is the set of path components of $\Omega^n X$. We index these components by the element in $\pi_n(X, *)$ to which a point corresponds under taking adjoint. Recall

LEMMA 3.4. *Any two path components* $\Omega^n_\alpha(X)$ *and* $\Omega^n_\beta(X)$ *of* $\Omega^n X$ *are homotopy equivalent.*

COROLLARY 3.5. *The fiber* F_n *in* 3.1 *is equal to* $\Omega^n_{+1}S^n \cup \Omega^n_{-1}S^n$, *the fiber* SF_n *of* $E: SG_{n+1} \to S^n$ *is equal to* $\Omega^n_{+1}S^n$ *and through dimension* $n-2$, $F_n \simeq G_n$, $SF_n \simeq SG_n$.

Proof. $\pi_n(S^n) = Z$ and a map $f: S^n \to S^n$ is a homotopy equivalence if and only if it has degree ± 1. The homotopy equivalences follow on looking at the homotopy exact sequences of the fibrations E and using that F and G are CW complexes.

Now consider the inclusion of the orthogonal group O_{n+1} in G_{n+1} induced from the usual action of O_{n+1} on S^n. The restriction of E to O_{n+1} is a fibration with fiber O_n and we have the diagram of fibrations

3.6
$$\begin{array}{ccccc} O_n & \longrightarrow & O_{n+1} & \xrightarrow{E} & S^n \\ \downarrow{\scriptstyle J} & & \downarrow{\scriptstyle J} & & \| \\ F_n & \longrightarrow & G_{n+1} & \xrightarrow{E} & S_n \end{array}$$

If we identify $\pi_i(F_n)$ with $\pi_{i+n}(S^n)$ as above, J_* becomes a map $J_*: \pi_i(O_n) \to \pi_{i+n}(S^n)$ which can be identified with the J-homomorphism of G. W. Whitehead [145]. Again from 3.6 we see on passing to the homotopy exact sequence of the fibrations that the boundary maps ∂ factor as

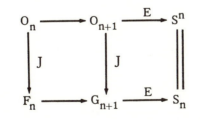

$$\begin{array}{ccccc}
\pi_i(S^n) & \xrightarrow{\ \partial\ } & \pi_{i-1}(O_n) & & \\
 & & \downarrow & \searrow^{J_*} & \\
 & & \pi_{i-1}(F_n) & \xrightarrow{\ \cong\ } & \pi_{i+n-1}(S^n)
\end{array}$$

We now consider the suspension map $i: G_{n+1} \to G_{n+2}$ defined in the beginning of this chapter. Let r be the composition,

$$r: F_n \to G_{n+1} \xrightarrow{i} F_{n+1} \, .$$

We record the following obvious result.

LEMMA 3.7. *Let* $k_n : S^n \to \Omega S^{n+1}$ *be adjoint to the identity* $S^1 \wedge S^n \to S^{n+1}$. *Then* $r: F_n \to F_{n+1}$ *is homotopic to* $\Omega^n k_n : \Omega^n S^n \to \Omega^{n+1} S^{n+1}$ *restricted to the* ± 1 *components.*

COROLLARY 3.8. *The induced homomorphism* $r_* : \pi_i(F_n) \to \pi_i(F_{n+1})$ *is an isomorphism for* $i < n-1$. *In particular*

$$\lim_{n\to\infty} \Omega^n_{\pm 1} S^n \simeq G \quad \textit{and} \quad \lim_{n\to\infty} \Omega^n_{+1} S^n \simeq SG \, ,$$

so $\pi_i(G) = \pi_i(SG) = \lim\limits_{k\to\infty} \pi_{i+k}(S^k)$.

Proof. This is just a restatement of the well-known fact that $k: S^n \to \Omega S^{n+1}$ is a homotopy equivalence in dimensions less than $2n-1$. (Indeed ΩS^{n+1} has the homotopy type of $S^n \cup e^{2n} \cup e^{3n} \cup \cdots$ via the reduced join construction [60] and k embeds S^n as the bottom cell.)

B. *The space* $Q(S^0)$ *and its structure*

If X is a space with base point the natural inclusions $\Omega^n S^n(X) \to \Omega^{n+1} S^{n+1}(X)$ on passing to the limit define the space $Q(X) = \lim\limits_{n\to\infty} \Omega^n S^n(X)$.

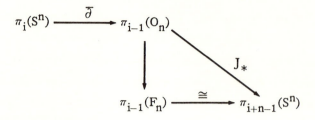

$Q(X)$ is a functor of X and $\pi_i(Q(X)) = \pi_i^s(X)$ is the i'th stable homotopy group of X. Dyer and Lashof in [50] were the first authors to study the structure of $Q(X)$. In the initial version of [50] but left out in the printed version they gave a geometric construction $C(X)$, an inclusion $i : C(X) \to Q(X)$ and showed that i is a weak homotopy equivalence for *connected* X,

$$C(X) = \coprod_{n \geq 1} E\Sigma_n \times_{\Sigma_n} X^n/\approx .$$

Here Σ_n is the symmetric group on n letters, $E\Sigma_n$ the universal covering space for $B\Sigma_n$ and Σ_n acts on X^n by permutation of coordinates. The equivalence relation $\approx$ identifies points of $E\Sigma_n \times_{\Sigma_n} F^1(X^n)$ with points of $E\Sigma_{n-1} \times_{\Sigma_{n-1}} X^{n-1}$ where $F^1(X^n) \subset X^n$ is the subset with at least one coordinate equal to $*$. Specifically, there are n maps $\tilde{\partial}_i : \Sigma_n \to \Sigma_{n-1}$ associated with the n monotone increasing embeddings of $\partial_i : \{1, \cdots, n-1\} \to \{1, \cdots, n\}$. Let $\tilde{\partial}_i : E\Sigma_n \to E\Sigma_{n-1}$ be the induced maps*) and let $s_i : X^{n-1} \to X^n$ be the embedding whose i'th coordinate is $*$. Then $\approx$ is the equivalence relation generated by $(\tilde{\partial}_i w, x) \approx (w, s_i x)$. We refer the reader to May [84] for more details.

The construction was rediscovered by Barratt and Priddy in 1969 and at about the same time by D. Quillen. Their description led to a similar result in the case that X is not necessarily connected. In particular, we give the following fundamental description of $Q(S^0)$. First note that

---

*) The adjoint map corresponding to $\partial_i, \tilde{\partial}_i : \Sigma_n \to \Sigma_{n-1}$ is defined to make the following diagram commutative

$$\begin{array}{ccc} \{1, \cdots, n-1\} & \xrightarrow{\partial_i} & \{1, \cdots, n\} \\ \Big\downarrow \tilde{\partial}_i(a) & & \Big\downarrow a \\ \{1, \cdots, n-1\} & \xrightarrow{\partial_{a(i)}} & \{1, \cdots, n\} \end{array}$$

Thus $\tilde{\partial}_{a(i)}(\beta) \cdot \tilde{\partial}_i(a) = \tilde{\partial}_i(\beta a)$ and the induced mapping $\tilde{\partial}_i : E\Sigma_n \to E\Sigma_{n-1}$ is determined by

$$\tilde{\partial}_i[g_1 | \cdots | g_n]g = [\tilde{\partial}_{g_2 \cdots g_n g(i)}(g_1) | \cdots | \tilde{\partial}_{g(i)}(g_n)]\tilde{\partial}_i(g) .$$

$$C = C(S^0) = \coprod_{n \geq 0} B\Sigma_n, \quad (B\Sigma_0 = * \text{ is the base point})$$

has a natural associative H-space structure with unit $*$, induced from the usual inclusion $\Sigma_n \times \Sigma_m \to \Sigma_{n+m}$ upon applying B and using $B(\Sigma_n \times \Sigma_m) = B\Sigma_n \times B\Sigma_m$. Barratt-Priddy and Quillen show ([16], [112]),

THEOREM 3.9. (i) *There are natural inclusions* $i_n : B\Sigma_n \to Q(S^0)$ *inducing an* H-*map* $I : C(S^0) \to Q(S^0)$ *with* $BI : BC(S^0) \to Q(S^1)$ *a homotopy equivalence. In particular,* $\Omega BC(S^0) \simeq Q(S^0)$.

(ii) $H_*(\Omega BC(S^0)) \cong \mathbf{Z} \times \lim_{n \to \infty} H_*(B\Sigma_n)$ *where the limit is induced from the natural inclusions of* Σ_n *in* Σ_{n+1} *as those permutions leaving* $n+1$ *fixed.*

We refer the reader to Segal [120], [121] for a very elegant proof of 3.9.

The product in $Q(S^0)$ which is related to that in ΩBC is the "loop sum" which is given as the limit of the $\Omega^{n-1}(*) : (\Omega^n S^n)^2 \to \Omega^n S^n$ where $* : \Omega S^n \times \Omega S^n \to \Omega S^n$ is the loop sum. However, the product which is appropriate to G and SG is composition of maps. Of course, $\Omega^n S^n$ as a space also admits composition product as well as loop product. It is then natural to ask how this composition product is reflected in the space $C(S^0)$ and the maps i_n. This requires a different family of homomorphisms. Let

$$\psi_{n,m} : \Sigma_n \times \Sigma_m \to \Sigma_{nm}$$

be the product of permutations ($\psi_{n,m}(\sigma, \tau)$ acts on the nm points (i, j) coordinatewise $\psi(\sigma,\tau)(i,j) = (\sigma(i), \tau(j))$). This only determines $\psi_{n,m}$ up to inner conjugation: we require a linear ordering of the (i, j) and usually we take the lexiographic ordering.

Our cohomological calculations in the remainder of this section are based on 3.9 and the following somewhat more precise

THEOREM 3.10. *The following diagram homotopy commutes for each pair* (n, m)

$$\begin{array}{ccc} B\Sigma_n \times B\Sigma_m & \xrightarrow{B(\psi_{n,m})} & B\Sigma_{nm} \\ \Big\downarrow{\scriptstyle i_n \times i_m} & & \Big\downarrow{\scriptstyle i_{nm}} \\ Q(S^0) \times Q(S^0) & \xrightarrow{\circ} & Q(S^0) \end{array}$$

where $\circ$ *is the map induced on passing to the limit from the composition products.*

Thus, we can also use the spaces $B\Sigma_n$ to study the composition product in $Q(S^0)$ and by restriction, in SG. This program begins with the specification of the $H_*(B\Sigma_n)$ and ends with a fairly clear description of $H_*(BSG)$. The original results at the prime 2 were discovered in [90] where $H^*(BSG; Z/2)$ was computed. The Pontrjagin ring $H_*(BSG; Z/2)$ was treated in [73] whereas at odd primes the results are primarily due to May [86] and Tsuchiya [139].

In the sequel we only discuss mod. 2 cohomology results. The results at odd primes are similar and left to the reader (cf. [106]).

C. *Wreath products, transfer, and the Sylow 2-subgroups of* Σ_n

Let Π be a finite group and $\mathrm{Aut}(\Pi)$ its group of automorphisms. Then $\mathrm{Aut}(\Pi)$ acts as a group of homeomorphisms of $B\Pi$,

$$B\alpha(t_1, \cdots, t_n, g_1, \cdots, g_n) = (t_1, \cdots, t_n, \alpha(g_1), \cdots, \alpha(g_n))$$

and we have the well known

LEMMA 3.11. *If* a *is an inner automorphism of* Π $(a \in \Pi)$ *then* $B\alpha : B\Pi \to B\Pi$ *is homotopic to the identity.*

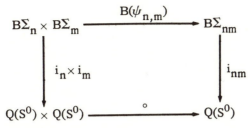

Proof. Define a homotopy $H^\alpha : I \times B\Pi \to B\Pi$ by

$$H^\alpha_t((t_1, \cdots, t_n, g_1, \cdots, g_n)) =$$
$$(t_1, \cdots, t_i, t, t_{i+1}, \cdots, t_n, \alpha(g_1), \cdots, \alpha(g_i), \alpha, g_{i+1}, \cdots, g_n)$$

where $t_{i+1} \le t \le t_i$. In view of the equivalence relations defining $B\Pi$ (compare 1.4) H^α_t is well defined when we recall $\alpha(g) = \alpha g \alpha^{-1}$, and since

$$(1, t_1, \cdots, t_n, g, g_1, \cdots, g_n) \approx (t_1, \cdots, t_n, g_1, \cdots, g_n)$$

$$(t_1, \cdots, t_n, 0, \alpha(g_1), \cdots, \alpha(g_n), g) \approx (t_1, \cdots, t_n, \alpha(g_1), \cdots, \alpha(g_n))$$

we see that H^α_t is indeed a homotopy of $B\alpha$ to the identity.

The subgroup $\mathrm{Int}(\Pi)$ of $\mathrm{Aut}(\Pi)$ consisting of inner automorphisms is normal in $\mathrm{Aut}(\Pi)$, and $\mathrm{Aut}(\Pi)/\mathrm{Int}(\Pi)$ acts on $H^*(B\Pi)$ in a (perhaps) non-trivial way. For example, $B(Z/2)^n = RP^\infty \times \cdots \times RP^\infty$ and $H^*(B(Z/2)^n) = P\{e_1, \cdots, e_n\}$. Also, $\mathrm{Aut}((Z/2)^n) = G\ell_n(Z/2)$ and this acts on $H^*(B(Z/2)^n)$ by $(B\theta)*(e_i) = \Sigma\theta_{ij} e_j$ where $\theta = {}^t(\theta_{ij})$.

COROLLARY 3.12. *Let* Γ *be a subgroup of* Π *and* $Bi : B\Gamma \to B\Pi$ *the associated inclusion of classifying spaces. Suppose* $N\Gamma$ *is the normalizer of* Γ *in* Π. *Then* $N\Gamma/\Gamma$ *acts on* $H^*(B\Gamma)$ *and the image of* $(Bi)^*$ *is contained in* $H^*(B\Gamma)^{N\Gamma/\Gamma}$, *the subgroup of invariants.*

Finally, we need the classical notion of the transfer associated with a finite covering space $E \to X$. It is the chain map

$$\mathrm{Tr}: C_*(X) \to C_*(E)$$

given by $\mathrm{Tr}(\sigma) = \Sigma\tilde{\sigma}$, where the sum is extended over all simplices (or singular simplices) of E covering σ. We see that $p \circ \mathrm{Tr}(x) = kx$ if $p : E \to X$ is a k-sheeted covering. Note that if $\Gamma \subset \Pi$ is a subgroup then $B\Gamma$ is homotopy equivalent to the orbit space $E\Pi/\Gamma$. In particular we have a covering $E\Pi/\Gamma \to B\Pi$ with $[\Pi:\Gamma]$ sheets. Summarizing

LEMMA 3.13. *Let* $i : \Gamma \subset \Pi$, *then there is a transfer mapping* $Tr: H_*(B\Gamma) \to H_*(B\Pi)$ *so that* $Tr \circ B_i^*$ *is multiplication by the index* $[\Pi : \Gamma]$.

(In particular, if Γ is a Sylow p-subgroup of Π then $(Bi)^* : H^*(B\Pi; Z/p) \to H^*(B\Gamma; Z/p)$ is a monomorphism as $[\Pi:\Gamma]$ is then prime to p.)

DEFINITION 3.14. Let $\Gamma \subset \Sigma_n$ and let H be any group. The wreath product $\Gamma \backslash H$ is defined to be the product $\Gamma \times H^n$ with multiplication specified by

$$(g', h'_1, \cdots, h'_n)(g, h_1, \cdots, h_n) = (g'g, h'_{g(1)}h_1, \cdots, h'_{g(n)}h_n) .$$

Note that H^n is normal in $\Gamma \backslash H$ with quotient group Γ. As a particular example $\Sigma_n \backslash \Sigma_m$ may be regarded as the set of permutations of pairs (i, j), $i = 1, \cdots, n$ and $j = 1, \cdots, m$ on defining

$$(g, h_1, \cdots, h_n)(i, j) = (g(i), h_i(j)) .$$

Using the lexiographic ordering as before, this gives an embedding

$$J_{n,m} : \Sigma_n \backslash \Sigma_m \to \Sigma_{nm} . \tag{3.15}$$

Also, there is an injection

$$1 \times \Delta : \Sigma_n \times \Sigma_m \to \Sigma_n \backslash \Sigma_m$$

where $(1 \times \Delta)(\sigma, \tau) = (\sigma, \tau, \cdots, \tau)$ and we have (compare 3.10)

$$J_{n,m} \circ (1 \times \Delta) = \psi_{n,m} .$$

LEMMA 3.16. *If* $\Gamma \subset \Sigma_n$ *and* H *is an arbitrary group then* $B(\Gamma \backslash H) \simeq E\Gamma \times_\Gamma (BH)^n$, *and the projection* $E\Gamma \times_\Gamma (BH)^n \to B\Gamma$ *is a fibering with fiber* $(BH)^n$ *and group* Γ.

Proof. We have a free action of $\Gamma\wr H$ on the acyclic space $E\Gamma\times EH\times\cdots\times EH$,

$$(w,v_1,\cdots,v_n)\cdot(g,e,\cdots,e) = (wg,v_{g(1)},\cdots,v_{g(n)})$$
$$(w,v_1,\cdots,v_n)\cdot(e,h_1,\cdots,h_n) = (w,v_1h_1,\cdots,v_nh_n)\,.$$

The orbit space, which is a model for $B(\Gamma\wr H)$ is clearly $E\Gamma\times_\Gamma(BH)^n$ as stated. The assertion about fibering is direct.

As a special case of 3.15 consider $J: Z/2\wr Z/2\to\Sigma_4$. Checking orders we see that J embeds the wreath product as a Sylow 2-subgroup of Σ_4. This may be iterated and we get an embedding

3.17 $$J:\underbrace{Z/2\wr\cdots\wr Z/2}_{n\text{ times}}\to Z/2\wr\Sigma_{2^{n-1}}\to\Sigma_{2^n}\,.$$

The 2-order of $n!$ is $\nu_2(n!)=n-\alpha(n)$ where $\alpha(n)$ is the number of terms in the dyadic expansion of n. Hence the 2-order of Σ_{2^n} is 2^n-1 and J embeds the iterated wreath product as a Sylow 2-subgroup. More genenerally, write m in its 2-adic expansion $m=2^{i_1}+\cdots+2^{i_j}$, $i_1<i_2<\cdots<i_j$. Then $\Sigma_{2^{i_1}}\times\cdots\times\Sigma_{2^{i_r}}\subset\Sigma_m$ is a subgroup of odd index and a Sylow 2-subgroup of Σ_m is then the product of the Sylow 2-subgroups of the factors (given in 3.17).

D. *A detecting family for the Sylow 2-subgroups of* Σ_n

DEFINITION 3.18. A (mod 2) detecting family of subgroups of Γ is a set of proper subgroups $\Gamma_1,\cdots,\Gamma_r$ so that

$$\{0\} = \bigcap_{j=1}^{r}\operatorname{Ker}(Bi_j^*: H^*(B\Gamma;Z/2)\to H^*(B\Gamma_j;Z/2))\,.$$

We begin by analyzing the cohomology with $Z/2$ coefficients of the "quadratic" construction $E\Sigma_2\times_{\Sigma_2}X\times X$ for any space X. Let Λ be the group ring $\Lambda=Z/2[\Sigma_2]$. The Eilenberg-Zilber map (see e.g.

MacLane [80]) implies a natural chain homotopy equivalence

$$C_*(E\Sigma_2) \otimes_\Lambda C_*(X) \otimes C_*(X) \xrightarrow{\simeq} C_*(E\Sigma_2 \times_{\Sigma_2} X \times X)\,.$$

Since $E\Sigma_2$ is a contractible Σ_2-free space its chain complex is a free Λ-resolution of $Z/2$ and hence homotopy equivalent to the standard resolution W,

3.19 $$W_n = \Lambda e_n\,, \qquad \partial(e_n) = (1+T)e_{n-1}$$

where $T \in \Sigma_2$ is the generator. Hence

$$W \otimes_\Lambda C_*(X) \otimes C_*(X) \simeq C_*(E\Sigma_2 \times_{\Sigma_2} X \times X)\,.$$

LEMMA 3.20. *Considering* $H_*(X)$ *as a chain complex with trivial differential there is a chain homotopy equivalence*

$$W \otimes_\Lambda C_*(X) \otimes C_*(X) \simeq W \otimes_\Lambda H_*(X) \otimes H_*(X)\,.$$

In particular $H_*(E\Sigma_2 \times_{\Sigma_2} X \times X)$ *is generated by elements* $e_i \otimes a \otimes a$ *and* $e_0 \otimes a \otimes b$ *for* $a, b \in H_*(X)$.

Proof. Associated to the free Λ chain complex $W \otimes C_*(X) \otimes C_*(X)$ there is a spectral sequence (see e.g. [80])

$$E^2_{*,*} = H_*(\Sigma_2; H_*(W \otimes C_*(X) \otimes C_*(X))) \Rightarrow H_*(W \otimes_\Lambda C_*(X) \otimes C_*(X))$$

and similarly for $C_*(X)$ replaced with $H_*(X)$. Let $f\colon C_*(X) \to H_*(X)$ be any map inducing isomorphisms on homology and consider

$$1 \otimes f \otimes f\colon W \otimes C_*(X) \otimes C_*(X) \to W \otimes H_*(X) \otimes H_*(X)\,.$$

This induces an isomorphism of the E^2-terms by the Künneth theorem and hence an isomorphism of the E^∞-terms. Finally, it is direct to calculate the homology of $W \otimes_\Lambda H_*(X) \otimes H_*(X)$.

Let H be any group. There are two important subgroups of the wreath product $\Sigma_2 \backslash H$, the normal subgroup $H \times H$ and the diagonal subgroup

$$\Sigma_2 \times H \xrightarrow{1\times\Delta} \Sigma_2 \backslash H .$$

To obtain $B(1\times\Delta)^*$ we recall the definition of the Steenrod squares given in [130]. A class $x \in H^n(X)$ induces a cocycle

$$x \otimes x = 1 \otimes x \otimes x \in \mathrm{Hom}(W \otimes_\Lambda H_*(X) \otimes H_*(X); Z/2)$$

hence from 3.20 a cohomology class again called $x \otimes x$ in $H^{2n}(E\Sigma_2 \times_{\Sigma_2} X^2; Z/2)$. Then the Steenrod squares are given by the formula

3.21 $$B(1\times\Delta)^*(x\otimes x) = \sum_{i=0}^{n} e^i \otimes Sq^{n-i}(x) .$$

LEMMA 3.22. *The subgroups* $\Sigma_2 \times H$ *and* $H \times H$ *form a detecting family for* $\Sigma_2 \backslash H$.

Proof. Let $e \in H^1(B(\Sigma_2 \backslash H))$ be the pull back of the generator under the projection $B(\Sigma_2 \backslash H) \to B\Sigma_2$. Combining 3.16 with 3.20 (dualized to cohomology) we see that the kernel of

$$H^*(B(\Sigma_2 \backslash H)) \to H^*(BH \times BH)$$

is generated by elements of the form $e^j \cup (x \otimes x)$ with $j > 0$ and $x \in H^*(BH)$. But 3.21 implies

$$B(1\times\Delta)^*(e^j \cup (x\otimes x)) = e^{j+n} \otimes x + \text{other terms.}$$

This completes the proof.

There is a very important detecting subgroup $V_n \subset \Sigma_{2^n}$ which we shall now describe. $V_n = Z/2 \times \cdots \times Z/2$ (n factors) with embedding

$$I: Z/2 \times \cdots \times Z/2 \xrightarrow{\bar{\Delta}} Z/2 \wr \cdots \wr Z/2 \xrightarrow{J} \Sigma_{2^n}$$

where $\bar{\Delta} = 1 \times \Delta_1 \times \Delta_2 \times \cdots \times \Delta_{n-1}$, $\Delta_i(x) = (x, \cdots, x) \in (Z/2)^{2^i}$ and J is the embedding of 3.17.

THEOREM 3.23. *The subgroups* $\Sigma_{2^{n-1}} \times \Sigma_{2^{n-1}}$ *and* V_n *are detecting for* Σ_{2^n}.

Proof. For $n = 2$ this is the statement of 3.22 and the fact that $Z/2 \wr Z/2$ is a Sylow 2-subgroup of Σ_4. Assume the result for $n-1$. The index of $Z/2 \wr \Sigma_{2^{n-1}}$ in Σ_{2^n} is odd so $Z/2 \wr \Sigma_{2^{n-1}}$ detects Σ_{2^n} by 3.13; hence $Z/2 \times \Sigma_{2^{n-1}}$ and $\Sigma_{2^{n-1}} \times \Sigma_{2^{n-1}}$ detect Σ_{2^n}. It follows from the inductive hypothesis that V_n, $Z/2 \times \Sigma_{2^{n-2}} \times \Sigma_{2^{n-2}}$ and $\Sigma_{2^{n-1}} \times \Sigma_{2^{n-1}}$ detect Σ_{2^n}.*) It remains to note that $Z/2 \times \Sigma_{2^{n-2}} \times \Sigma_{2^{n-2}} \subset \Sigma_{2^n}$ is conjugate to a subgroup of $\Sigma_{2^{n-1}} \times \Sigma_{2^{n-1}}$. This completes the proof.

E. *The image of* $H^*(B\Sigma_n)$ *in the cohomology of the detecting groups*

We next compute the image of $H^*(B\Sigma_{2^n})$ in $H^*(BV_n) = P\{e_1, \cdots, e_n\}$. An upper bound is found using 3.12 and a result of L. E. Dickson from 1911 [44]. The Stiefel-Whitney classes of the permutation representation $\Sigma_{2^n} \subset O_{2^n}(\mathbf{R})$ give a lower bound.

Consider Σ_{2^n} as the automorphism group of the underlying set of the $Z/2$-vector space $Z/2 \oplus \cdots \oplus Z/2$ (n summands). Then $V_n \subset \Sigma_{2^n}$ is the set of translations and its normalizer the group of affine transformations, $\mathrm{Aff}_n(Z/2)$. Consequently, $N(V_n)/V_n = G\ell_n(Z/2)$ the general linear group

---

*)We are using here that V_n is conjugate to $V_n' \subset \Sigma_{2^n}$ inductively defined by $V_n' = Z/2 \times V_{n-1}' \subset Z/2 \times \Sigma_{2^{n-1}} \subset Z/2 \wr \Sigma_{2^{n-1}} \subset \Sigma_{2^n}$. Indeed, if one consistently uses the lexiographic ordering to embed the wreath products then $V_n' = V_n$.

and the image of $H^*(B\Sigma_{2^n})$ is contained in the ring of invariants

$$H^*(BV_n)^{G\ell_n(\mathbf{Z}/2)} = P\{e_1, \cdots, e_n\}^{G\ell_n(\mathbf{Z}/2)} .$$

Let $\hat{D}_i$ be the matrix

$$\hat{D}_i = \begin{pmatrix} e_1 & e_2 & \cdots & e_n \\ e_1^2 & e_2^2 & \cdots & e_n^2 \\ \vdots & \vdots & & \vdots \\ e_1^{2^{i-1}} & e_2^{2^{i-1}} & \cdots & e_n^{2^{i-1}} \\ e_1^{2^{i+1}} & e_2^{2^{i+1}} & \cdots & e_n^{2^{i+1}} \\ \vdots & \vdots & & \vdots \\ e_1^{2^n} & e_2^{2^n} & \cdots & e_n^{2^n} \end{pmatrix}$$

and let $D_i = \det(\hat{D}_i) \in P\{e_1, \cdots, e_n\}^{G\ell_n(\mathbf{Z}/2)}$. The following theorem is proved in [44].

THEOREM 3.24 (Dickson). *For* $i < n$ D_n *divides* D_i *and* $P\{e_1, \cdots, e_n\}^{G\ell_n(\mathbf{Z}/2)}$ *is the polynomial subalgebra with generators*

$$D_{n-1}/D_n, D_{n-2}/D_n, \cdots, D_1/D_n, D_n$$

(*degree* $(D_i/D_n) = 2^n - 2^i$, *degree* $D_n = 2^n - 1$).

Next we have the real representation $\gamma_n : V_n \to \Sigma_{2^n} \xrightarrow{P_n} O_{2^n}(\mathbf{R})$ where P_n is the permutation representation (embedding Σ_{2^n} as the permutation group of the 2^n coordinates in $\mathbf{R}^{2^n}$). γ_n is the regular representation of V_n and from the theory of representations of finite groups the character of γ_n is the sum of all the irreducible characters of V_n. Now, the irreducible representations of $(Z/2)^n$ are all real, one dimensional and can be identified with the elements in $\mathrm{Hom}(V_n, Z/2)$ since $Z/2 = O_1$. Identifying $\mathrm{Hom}(V_n, Z/2)$ with $H^1(BV_n; Z/2)$ we see that the total Stiefel-Whitney class*) of the representation f is $1+f$. Hence the total Stiefel-Whitney class of γ_n is

$$W(\gamma_n) = \prod_{f \in H^1(BV_n)} (1+f) .$$

LEMMA 3.25 (E. H. Moore [104]). $D_n = \prod_{f \neq 0} f, \; f \in H^1(BV_n)$.

Proof. From the definition we see that e_1 divides D_n and as D_n is invariant under $G\ell_n(Z/2)$ each non-zero $f \in H^1(BV_n)$ must divide D_n. Moreover, by inspection $D_n \neq 0$. Comparing degrees the result follows.

LEMMA 3.26. *The total Stiefel-Whitney class of* γ_n *is*

$$W(\gamma_n) = 1 + D_{n-1}/D_n + D_{n-2}/D_n + \cdots + D_1/D_n + D_n$$

Proof. From 3.25 we have that $w_{2^n-1}(\gamma_n) = D_n$. In degree $\leq 2^n - 1$ the only elements in $P\{e_1, \cdots, e_n\}^{G\ell_n(Z/2)}$ are D_n and the D_i/D_n. Then we calculate from the definition

3.27 $$Sq^{2^i}(D_{i+1}/D_n) = D_i/D_n \qquad (D_0/D_n = D_n) .$$

Finally, we recall the Wu formula for the action of the Sq^i on the Stiefel-Whitney class w_j,

*) The i'th Stiefel-Whitney class of a representation $\gamma: \pi \to O_n$ is defined as $w_i(\gamma) = (B\gamma)^*(w_i)$, cf. 1.13.

$$Sq^i(w_j) = \binom{j-1}{i} w_{i+j} + \text{decomposable terms} .$$

These formulae, together with the fact $w_{2^n-1}(\gamma_n) = D_n$ inductively give the result.

COROLLARY 3.28. *The image of* $(BI)^*: H^*(B\Sigma_{2^n}) \to H^*(BV_n)$ *is precisely the ring of invariants* $H^*(BV_n)^{G\ell_n(\mathbf{Z}/2)}$.

We now consider the images of $w_i(P_n)$ in $H^*(B\Sigma_{2^{n-1}} \times B\Sigma_{2^{n-1}})$ under the inclusion $j: B\Sigma_{2^{n-1}} \times B\Sigma_{2^{n-1}} \to B\Sigma_{2^n}$. First, the restriction of P_n to $\Sigma_{2^{n-1}} \times \Sigma_{2^{n-1}}$ is the (exterior) sum $P_{n-1} \times P_{n-1}$. Second, $P_n = \varepsilon \oplus P'_n$ where ε is the trivial 1-dimensional representation. Indeed, if e_1 is any non-zero vector of the representation space $\mathbf{R}^{2^n}$ of P_n then $e = \Sigma\{g \cdot e_1 | g \in \Sigma_{2^n}\}$ is non-zero and invariant under the action of Σ_{2^n}. Hence $P_n | \Sigma_{2^{n-1}} \times \Sigma_{2^{n-1}} = 2\varepsilon \oplus P''$ and in particular the induced bundle

$$B\Sigma_{2^{n-1}} \times B\Sigma_{2^{n-1}} \xrightarrow{j} B\Sigma_{2^n} \longrightarrow BO_{2^n}$$

has two sections, so

3.29 $$j^*(w_{2^n-1}(P_n)) = 0 .$$

Now using 3.29 and the inclusions

$$V_{n-1} \times V_{n-1} \subset \Sigma_{2^{n-1}} \times \Sigma_{2^{n-1}} \xrightarrow{j} \Sigma_{2^n}$$

we see that j^* injects the subpolynomial algebra

$P\{w_{2^{n-1}}(P_n), \cdots, w_{2^n-2}(P_n)\}$ into $H^*(B\Sigma_{2^{n-1}} \times B\Sigma_{2^{n-1}})$. From 3.23 we then have

COROLLARY 3.30. *The kernel of* j^* *is exactly the ideal* $P\{w_{2^n-1}(P_n), \cdots, w_{2^n-2}(P_n)\} \cdot w_{2^n-1}(P_n)$.

EXAMPLE 3.31. We use the above results to determine $H*(B\Sigma_4)$. In fact we claim

$$H^*(B\Sigma_4) = P\{w_1(P), w_2(P), w_3(P)\}/< w_1(P) \cdot w_3(P) = 0> .$$

Indeed the detecting groups are $\Sigma_2 \times \Sigma_2$ and V_2. The normalizer of $\Sigma_2 \times \Sigma_2$ in Σ_4 is $\Sigma_2 \wr \Sigma_2$, hence the image of $H^*(B\Sigma_4)$ in $H^*(B\Sigma_2 \times B\Sigma_2)$ is contained in $P\{e_1, e_2\}^{\Sigma_2}$ where the generator T of Σ_2 switches e_1 and e_2. But $P\{e_1, e_2\}^{\Sigma_2} = P\{e_1 + e_2, e_1 e_2\}$, and $e_1 + e_2 = j^*(w_1(P))$, $e_1 e_2 = j^*(w_2(P))$. We have already shown that the image of $H^*(B\Sigma_4)$ in $H(BV_2)$ is

$$\begin{aligned} P\{D_1/D_2, D_2\} &= P\{e_1^2 + e_1 e_2 + e_2^2, e_1^2 e_2 + e_1 e_2^2\} \\ &= P\{w_2(P), w_3(P)\} \end{aligned}$$

and the claim follows.

The situation above is atypical in the sense that it is not true that $H^*(B\Sigma_{2^n})$ for $n > 2$ is generated by the $w_i(P)$. However, it does reflect a viable though complex procedure for inductively determining the cohomology of the symmetric groups.

In order to describe the answer most efficiently we switch to homology. Dual to 3.28 we have an embedding

$$(BI)_*: H_*(BV_n)_{G\ell_n(\mathbf{Z}/2)} \to H_*(B\Sigma_{2^n})$$

where $H_*(BV_n)_{G\ell_n(\mathbf{Z}/2)}$ denotes the coinvariants. Let $E_{(i_1, \cdots, i_n)}$ be the basis dual to the monomial basis for $H^*(BV_n)$ with $E_{(i_1, \cdots, i_n)}$ dual to $w_{2^n-1}^{i_1} \cdot \cdots \cdot w_{2^n-1}^{i_n}$. We will also write $E_{(i_1, \cdots, i_n)}$ for its image under $(BI)_*$ in $H_*(B\Sigma_{2^n})$.

Recall the space $C(S^0) = \coprod B\Sigma_n$ defined in 3.9. We now can state

THEOREM 3.32. $H_*(C(S^0)) = \mathbf{Z}_+ \times P\{E_{(i_1, \cdots, i_n)} | n \geq 1, i_1 > 0\}$.

The proof outlined here uses concepts which may be unfamiliar to the reader. For this reason a second detailed proof using more familiar concepts but harder combinatorial calculations is included as Chapter 3.G.

Let $SP^n(X)$ be the n'th symmetric product consisting of all unordered n-tuples of points $x_1, \cdots, x_n$ from X. It is the quotient of X^n by the permutation action of Σ_n, and is topologized by giving it the quotient topology. We need four results on symmetric products. From [49] we have

LEMMA 3.33 (Dold-Thom). *For a finite* CW *complex* X, $SP^\infty(X) = \lim_{n \to \infty} SP^n(X)$ $\simeq \prod K(H_i(X; \mathbf{Z}), i)$, *the product of Eilenberg-MacLane spaces.*

(The idea of the proof was to show that the functor SP^∞ converts cofibrations into quasi-fiberings, and hence that $\pi_*(SP^\infty(X))$ becomes a homology theory on X. Now, $SP^n(\mathbf{C}) = \mathbf{C}^n$ by assigning to an unordered n-tuple $a_1, \cdots, a_n$ the coefficients in the polynomial $\prod(t-a_i)$, and similarly $SP^n(\mathbf{C}^\times) = \mathbf{C}^\times \times \mathbf{C}^{n-1}(\mathbf{C}^\times = \mathbf{C} - \{0\})$, so $SP^\infty(S^1) \simeq S^1 = K(\mathbf{Z}, 1)$. Then 3.33 follows by an inductive argument over the cells of X.)

LEMMA 3.34 (Steenrod [132]). $\tilde{H}_*(SP^n(X)) = \bigoplus_{m \leq n} H_*(SP^m(X), SP^{m-1}(X))$. *In particular* $SP^n(X) \to SP^{n+1}(X)$ *induces an injection in homology onto a direct summand.*

(This is proved by semi-simplicial methods by constructing a chain retraction $C_*(SP^m(X), SP^{m-1}(X)) \to C*(SP^m(X))$ over the projection.)

LEMMA 3.35. *For* $i \leq 2m-2$, $H_i(B\Sigma_n) = H^{2nm-i}(SP^n(S^{2m}))$.

(Σ_n acts freely on cells in $(S^{2m})^n$ of dimension $> 2m(n-1)$ (and $< 2mn$) and trivially on the top cell. Further, $H^i((S^{2m})^n; Z) = 0$ in this range except $H^{2mn}((S^{2m})^n; Z) = Z$ with trivial Σ_n action. The cochains form a Σ_n-free resolution of Z in the same range.)

The H-structure in $C(S^0)$ is induced from the usual inclusions $Bi: B\Sigma_n \times B\Sigma_\ell = B(\Sigma_n \times \Sigma_\ell) \to B\Sigma_{n+\ell}$ and in homology we have

LEMMA 3.36. *The composition*

$$H^{2mn-i}(SP^n(S^{2m})) \otimes H^{2m\ell-j}(SP^\ell(S^{2m})) \xrightarrow{S\otimes S}$$
$$H^{2mn-i}(SP^{n+\ell}(S^{2m})) \otimes H^{2m\ell-j}(SP^{n+\ell}(S^{2m})) \xrightarrow{\text{cup-prod.}}$$
$$H^{2m(\ell+n)-(i+j)}(SP^{n+\ell}(S^{2m}))$$

in our range can be identified with the inclusion $(Bi)_*: H_*(B\Sigma_n) \otimes H_*(B\Sigma_\ell) \to H_*(B\Sigma_{n+\ell})$, *where* S *is the Steenrod retraction of* 3.34.

(The argument is similar to 3.35 and uses the fact that in $(S^{2m})^{\ell+n}$ one has

$$(e^{2m})^{\otimes n} \otimes 1^{\otimes \ell} \cup 1^{\otimes n} \otimes (e^{2m})^{\otimes \ell} = (e^{2m})^{\otimes(\ell+n)},$$

the top dimensional generating class.)

Proof of 3.32. The elements $E_{(i_1,\cdots,i_n)} \in H_*(B\Sigma_{2^n}) \subset H_*(C(S^0))$ and $[s] \in H_0(B\Sigma_s)$ generate $H_*(C(S^0))$. This follows from 3.23, 3.28 and 3.30 once we note that $E_{(i_1,\cdots,i_n)} * [s]$ is the image of $E_{(i_1,\cdots,i_n)}$ in $H_*(B\Sigma_{2^n+s})$ under the inclusion $H_*(B\Sigma_{2^n}) \subset H_*(B\Sigma_{2^n+s})$. But $E_{(0,i_2,\cdots,i_n)} = E_{(i_2,\cdots,i_n)} * E_{(i_2,\cdots,i_n)}$, so $H_*(C(S^0))$ is a quotient of the stated algebra and we must check for relations. It is well known that the cohomology of an Eilenberg-MacLane space $K(Z,2m)$ is a polynomial algebra in given generators (certain admissible monomials in the Steenrod squares). Now $SP^\infty(S^{2m}) \simeq K(Z,2m)$ and the result follows from 3.30 (which shows that $E_{(i_1,\cdots,i_n)}$ is indecomposable for $i_1 > 0$) and 3.36.

To recover the homology of a single Σ_n from 3.32 we proceed as follows. We assign to $E_{(i_1,\cdots,i_n)} \in H_*(B\Sigma_{2^n})$ the weight 2^n and to $[s] \in H_0(B\Sigma_s)$ the weight s. Then, the weight of a product of terms above will be the sum of the weights of the individual terms. By 3.32 this is well defined in $H_*(C(S^0))$ on monomials. Then $H_*(B\Sigma_n)$ is the $Z/2$ vector space generated by the monomials of weight exactly n (cf. [90], [107]).

EXAMPLE 3.37. (a) $H_*(B\Sigma_3) = H_*(B\Sigma_2) * [1]$.
(b) $H_*(B\Sigma_6) = H_*(B\Sigma_4) \otimes H_*(B\Sigma_2)/R$, where R is the set of relations:

$$(E_{i_1} * E_{i_2}) \otimes E_{i_3} = (E_{i_1} * E_{i_3}) \otimes E_{i_2} = (E_{i_2} * E_{i_3}) \otimes E_{i_1}$$
$$(E_{i_1} * [2]) \otimes E_{i_2} = (E_{i_1} * E_{i_2}) \otimes [2].$$

F. *The homology of* $Q(S^0)$ *and* SG

We are now ready to describe the homology ring of $Q(S^0)$ under loop sum and of SG under composition product.

From 3.9 (cf. footnote on page 12) $Q(S^0)$ is the group completion of $C(S^0)$. Since

$$H_*(\text{Group compl. } X) = \mathrm{Gp}(\pi_0(X)) \times \varinjlim H_*(X)$$

where $\varinjlim$ is taken over the component inclusions and $\mathrm{Gp}(\pi_0(X))$ is the group associated to the monoid of path components (see e.g. [120]) 3.32 gives us

THEOREM 3.38.

$$H_*(Q(S^0)) = P\{E_{(i_1,\dots,i_n)} | n \geq 1, i_1 > 0\} \otimes Z/2[Z]$$

where $k \in Z$ *represents the generator of* $H_0(Q_k(S^0))$ *and where we have suppressed the injection* $(BI)_*$. *The degree of* $E_{(i_1,\dots,i_n)}$ *is* $(2^n-1)i_1 + \dots + 2^{n-1} i_n$.

REMARK 3.39. From the description of the $E_{(i_1,\dots,i_n)}$ (given in the paragraph preceding 3.32) it is direct to see that their coproduct is given by

$$\psi(E_{(i_1,\cdots,i_n)}) = \sum E_{(j_1,\cdots,j_n)} \otimes E_{(k_1,\cdots,k_n)}$$

$(j_\nu + k_\nu = i_\nu)$. In particular, the cohomology ring $H^*(Q(S^0))$ is again a polynomial algebra with generators in one to one correspondence with the generators for $H_*(Q(S^0))$. More precisely, we can define $F_I \in H^*(Q(S^0))$ to be dual to E_I if, in $I = (i_1, \cdots, i_n)$ some entry i_j is odd. Otherwise, if $I = 2^r J$, $r > 0$ (and r maximal) we define F_I to be dual to $E_J^{2^r}$. (Dual here means dual with respect to the monomial basis E_I, hence F_I is primitive if I contains an odd entry). We have

$$H^*(Q(S^0)) = P\{\cdots, F_I, \cdots\} \otimes Z/2[Z] .$$

The formula 3.27 allows one to calculate the action of the Steenrod algebra when we also adjoin the result

$$Sq^{2^{n-1}}(D_n) = D_{n-1} = (D_{n-1}/D_n) \cdot D_n$$

which is obvious from the defining relation. For example, through dimension 7 we have

Dimension	1	2	3	4	5	6	7
Generators	$F_{(1)}$	$F_{(2)}$	$F_{(3)}$ $F_{(1,0)}$	$F_{(4)}$	$F_{(5)}$ $F_{(1,1)}$	$F_{(6)}$ $F_{(2,0)}$	$F_{(7)}$ $F_{(1,2)}$ $F_{(1,0,0)}$

The action of the Steenrod algebra is described by the formulae:

$$Sq^1(F_{(2)}) = F_{(3)} + F_{(1)}^3 + F_{(1,0)}$$

$$Sq^1(F_{(3)}) = F_{(1)}^4,\ Sq^1(F_{(1,0)}) = 0$$

$$Sq^2(F_{(3)}) = F_{(5)},\ Sq^2(F_{(1,0)}) = F_{(1,1)}$$

$$Sq^1(F_{(4)}) = F_{(5)} + F_{(3)}F_{(2)} + F_{(3)}F_{(1)}^2 + F_{(2)}F_{(1)}^3 + F_{(2)}F_{(1,0)} + F_{(1,1)}$$

$$Sq^2(F_{(4)}) = F_{(6)} + F_{(2)}^3 + F_{(1)}^3 F_{(3)} + F_{(1)}F_{(5)} + F_{(1)}^3 F_{(1,0)} + F_{(3)}F_{(1,0)} + F_{(2,0)} + F_{(1)}^6$$

$$Sq^1(F_{(5)}) = F_{(3)}^2$$

$$Sq^2(F_{(5)}) = 0$$

$$Sq^1(F_{(6)}) = F_{(1)}^4 \cdot F_{(3)} + F_{(1,2)}$$

$$Sq^1(F_{(2,0)}) = 0 .$$

With this as preparation we proceed to consider the composition product. The essential result is 3.10 and using it and 3.32 the calculation is in principle direct, but the formulae one obtains soon become unmanageable. The process is considerably simplified by the next lemma. Let $a, b \in H_*(Q(S^0))$, then the composition product will be denoted $a \circ b$ or just ab and the loop sum will be denoted $a * b$.

LEMMA 3.40. *Let* $a, b, c \in H_*(Q(S^0))$ *and write* $\Delta_*(c) = \Sigma c_i' \otimes c_i''$ *where* $\Delta : Q(S^0) \to Q(S^0) \times Q(S^0)$ *is the diagonal map, then*

(i) $$a \circ b = b \circ a$$

(ii) $$(a * b) \circ c = \Sigma (a \circ c_i') * (b \circ c_i'') .$$

Proof. (i) If a is represented in $H_*(\Omega^n S^n)$ and b in $H_*(\Omega^m S^m)$, then in $H_*(\Omega^{n+m} S^{n+m})$ a may be represented by a chain A carried by singular simplexes whose points give maps $f : (I^{n+m}, \partial) \to (I^{n+m}, \partial)$ which satisfy

$$f(t_1, \cdots, t_n, t_{n+1}, \cdots, t_{n+m}) = (f_1(t_1, \cdots, t_n), \cdots, f_n(t_1, \cdots, t_n), t_{n+1}, \cdots, t_{n+m}) .$$

Similarly, b may be represented by a chain B which is the identity on the first n coordinates,

$$g(t_1, \cdots, t_{n+m}) = (t_1, \cdots, t_n, g_1(t_{n+1}, \cdots, t_{n+m}), \cdots, g_m(t_{n+1}, \cdots, t_{n+m})) .$$

Clearly, $A \circ B = B \circ A$ and represents $a \circ b = b \circ a$. (Note that 3.8 shows that for each $a \in H_*(Q(S^0))$ there is an n so $a \in \text{Image}\,(H_*(\Omega^n S^n) \to H_*(Q(S^0)))$.

To prove (ii) we use a similar argument assuming to begin that $a * b$ affects only the first n coordinates and c only the last m so if $f * g$ belongs to a simplex of $A * B$ and h to a simplex of C we have $(f * g) \circ h = f \circ h * g \circ h$. Hence the diagram below homotopy commutes on finite subcomplexes

$$\begin{array}{ccccc}
Q(S^0)\times Q(S^0)\times Q(S^0) & & \xrightarrow{\;1\times 1\times\Delta\;} & & Q(S^0)^2\times Q(S^0)^2 \\
 & & & & \Big\downarrow {\scriptstyle 1\times T\times 1} \\
\Big\downarrow {\scriptstyle *\times 1} & & & & Q(S^0)^2\times Q(S^0)^2 \\
 & & & & \Big\downarrow {\scriptstyle (\circ)\times(\circ)} \\
Q(S^0)\times Q(S^0) & \xrightarrow{\;\circ\;} & Q(S^0) & \xleftarrow{\;*\;} & Q(S^0)
\end{array}$$

From this (ii) follows on passing to homology.

Lemma 3.40 allows us to reduce consideration to just the loop sum indecomposables, which are taken care of in 3.10 except for $[-1]$, the non-zero class in $H_0(Q_{-1}(S^0))$. In a loop space ΩY there is an involution $\chi : \Omega Y \to \Omega Y$ where $\chi(f)(t) = f(1-t)$, which serves as a homotopy inverse for the loop sum in the sense that the composite

3.41 $$\Omega Y \xrightarrow{\;\Delta\;} \Omega Y \times \Omega Y \xrightarrow{\;\chi\times 1\;} \Omega Y \times \Omega Y \xrightarrow{\;*\;} \Omega Y$$

is homotopic to the constant map $\Omega Y \to *$. The set of components of ΩY forms a group isomorphic to $\pi_1(Y)$ under loop sum. Let $[\ell] \in H_0(\Omega_\ell Y)$ denote the non-zero element in the component of $\ell \in \pi_1(Y)$. Then $\chi_*([\ell]) = [\chi(\ell)]$ and since $\chi(f * g) = \chi(g) * \chi(f)$ we get on passing to homology,

LEMMA 3.42. (i). $\chi_*(a * b) = (-1)^{|a|\,|b|}\chi_*(b) * \chi_*(a)$.
(ii) *If* $a \in H_*(\Omega_\ell Y)$, *where* $\Omega_\ell Y$ *is the* ℓ*-component of* ΩY *and*

$$\Delta_*(a) = a \otimes [\ell] + [\ell] \otimes a + \Sigma\, a_i' \otimes a_i''$$

with $\deg(a_i') > 0$, $\deg(a_i'') > 0$, *then*

$$\chi_*(a) * [\ell] = -\chi_*([\ell]) * a - \Sigma\, \chi_*(a_i') * a_i'' .$$

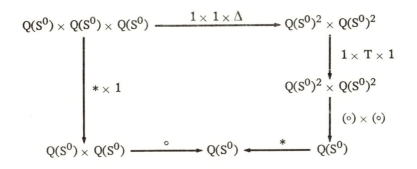

The connection between χ_* and the composition product in $Q(S^0)$ is given by

LEMMA 3.43. *For all* $a \in H_*(Q(S^0))$ *we have* $a \circ [-1] = \chi_*(a)$.

Proof. We suppose a represented in $H_*(\Omega^n S^n)$ for some n so that the points in the simplices of some chain representing a can be assumed to move only the first $(n-1)$ coordinates, as maps $(I^n, \partial) \to (S^n, *)$. Moreover, we can assume loop sum given on the n'th coordinate. Then $[-1]$ is represented by

$$\phi : (t_1, \cdots, t_n) \to (t_1, \cdots, t_{n-1}, 1-t_n)$$

and clearly $a \circ \phi = \chi(a)$.

With this preparation we can effectively calculate $H_*(SG)$ as a ring with respect to the composition product.

THEOREM 3.44. *The Pontrjagin ring of* SG *under the composition product is*

$$H_*(SG) = E\{\bar{e}_n | n \geq 1\} \otimes P\{\bar{e}_n * \bar{e}_n * [-1] | n \geq 1\} \otimes P\{\bar{E}_{(i_1, \cdots, i_n)} | i_1 > 0, n > 1\}$$

where $\bar{E}_{(i_1, \cdots, i_n)} = E_{(i_1, \cdots, i_n)} * [1-2^n]$ *and* $\bar{e}_n = E_{(n)} * [-1]$.

Proof. We introduce a filtration on $H_*(QS^0)$

$$F_i H_*(QS^0) = \{x | \ell(x) \geq i\}$$

where $\ell(E_{(i_1, \cdots, i_n)}) = 2^n$, $\ell([s]) = 0$, $\ell(x * y) = \ell(x) + \ell(y)$ and $\ell(x + y) = \min(\ell(x), \ell(y))$. Note from the remarks following the proof of 3.32 that each element in the image of

$$H_*(BV_n) \longrightarrow H_*(B\Sigma_{2^n}) \xrightarrow{(i_n)_*} H_*(QS^0)$$

has "length" equal to 2^n. The diagram

$$\begin{array}{ccc} V_n \times V_m & \xrightarrow{I \times I} & \Sigma_{2^n} \times \Sigma_{2^m} \\ \big\downarrow & & \big\downarrow \psi \\ V_{n+m} & \xrightarrow{I} & \Sigma_{2^{n+m}} \end{array}$$

shows that

$$\ell(E_{(i_1, \cdots, i_n)} \circ E_{(j_1, \cdots, j_m)}) = 2^{n+m} .$$

From 3.40(ii), 3.42(ii) and 3.43 we then have that

$$\bar{E}_{(i_1, \cdots, i_n)} \circ \bar{E}_{(j_1, \cdots, j_m)} = E_{(i_1, \cdots, i_n)} * E_{(j_1, \cdots, j_m)} * [1-2^n-2^m]$$

modulo elements of higher filtration provided $n+m > 2$. Thus, to complete the proof we must show that $\bar{e}_n \circ \bar{e}_n = 0$ or, in view of 3.40, 3.42 and 3.43 that

$$E_{(n)} * E_{(n)} = E_{(n)} \circ E_{(n)} .$$

But this is clear (by 3.11) once we remark that the two embeddings

$$Z/2 \xrightarrow{\Delta} Z/2 \times Z/2 \xrightarrow{\psi} \Sigma_4$$

$$Z/2 \xrightarrow{\Delta} Z/2 \times Z/2 \xrightarrow{*} \Sigma_4$$

map onto conjugate subgroups of Σ_4. This completes the proof.

Applying the Eilenberg-Moore spectral sequence (cf. the footnote to 1.5(e))

$$\mathrm{Ext}_{H_*(SG)}(Z/2, Z/2) \Longrightarrow H^*(BSG)$$

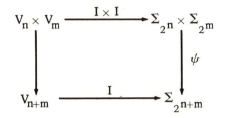

(or the Serre spectral sequence along with results of Browder [25], [28] on its structure for fiberings of H-spaces), we then obtain

THEOREM 3.45.

$$H^*(BSG) = P\{w_2, w_3, \cdots\} \otimes E\{\sigma(\bar{e}_n * \bar{e}_n * [-1])^* | n \geq 1\}$$
$$\otimes E\{\sigma(\bar{E}_{(i_1, \cdots, i_n)})^* | i_1 > 0, n > 1\}$$

where σ denotes the homology suspension, and $\sigma(\bar{E}_{(i_1,\cdots,i_n)})^*$ denotes a cohomology class dual to $\sigma(\bar{E}_{(i_1,\cdots,i_n)})$ etc.

(One can use the calculational results

(i) $\mathrm{Ext}_{A\otimes B}(Z/2, Z/2) = \mathrm{Ext}_A(Z/2, Z/2) \otimes \mathrm{Ext}_B(Z/2, Z/2)$

(ii) $\mathrm{Ext}_{E\{f\}}(Z/2, Z/2) = P\{[f]\}$,

(iii) $\mathrm{Ext}_{P\{f\}}(Z/2, Z/2) = E\{[f]\}$,

where $\deg[f] = 1 + \deg(f)$ to evaluate the E_2-term. Then we use naturality to check that differentials vanish. A key observation is that the piece $P\{[e_0], [e_1], \cdots, [e_n], \cdots\}$ must be identified with $P\{w_1, w_2, \cdots, w_{n+1}, \cdots\}$, the polynomial algebra on the universal Stiefel-Whitney classes.)

G. *The proof of Theorem 3.32*

The definition 3.21, of the Steenrod squares may be generalized using higher symmetric groups. A brief description follows, for further details we refer to [130].

A class $\Gamma_m(a) \in H^*(E\Sigma_m \times_{\Sigma_m} X^m)$ is defined for any class $a \in H^*(X)$. On the chain level $\Gamma_m(a)$ is $1 \otimes a \otimes \cdots \otimes a$, so $\Gamma_2(a)$ is the class we denoted $a \otimes a$ in 3.21. It has the following properties

3.46 (i) $$\Gamma_m(a \cup b) = \Gamma_m(a) \cup \Gamma_m(b) .$$

From the map $\Sigma_k \times \Sigma_{m-k} \to \Sigma_m$ we obtain

$$\phi_{k,m-k} : (E\Sigma_k \times_{\Sigma_k} X^k) \times (E\Sigma_{m-k} \times_{\Sigma_{m-k}} X^{m-k}) \to E\Sigma_m \times_{\Sigma_m} X^m$$

and

3.46 (ii) $$\phi^*_{k,m-k}(\Gamma_m(a)) = \Gamma_k(a) \otimes \Gamma_{m-k}(a) .$$

Let $J_{k,r} : \Sigma_k \wr \Sigma_r \to \Sigma_{kr}$ be the usual inclusion (cf. 3.15) and consider the composite

$$j_{k,r} : E\Sigma_k \times_{\Sigma_k} (E\Sigma_r \times_{\Sigma_r} X^r)^k = E(\Sigma_k \wr \Sigma_r) \times_{\Sigma_k \wr \Sigma_r} X^{kr} \to E\Sigma_{kr} \times_{\Sigma_{kr}} X^{kr} .$$

We have

3.46 (iii) $$j^*_{k,r}(\Gamma_{kr}(a)) = \Gamma_k(\Gamma_r(a)) .$$

Finally the construction is natural: Given $f : X \to Y$ this induces $E(f) : E\Sigma_m \times_{\Sigma_m} X^m \to E\Sigma_m \times_{\Sigma_m} Y^m$ and

3.46 (iv) $$E(f)^*(\Gamma_m(a)) = \Gamma_m(f^*(a)) .$$

Consider the commutative diagram (where $m = 2^{n-1}$).

3.47

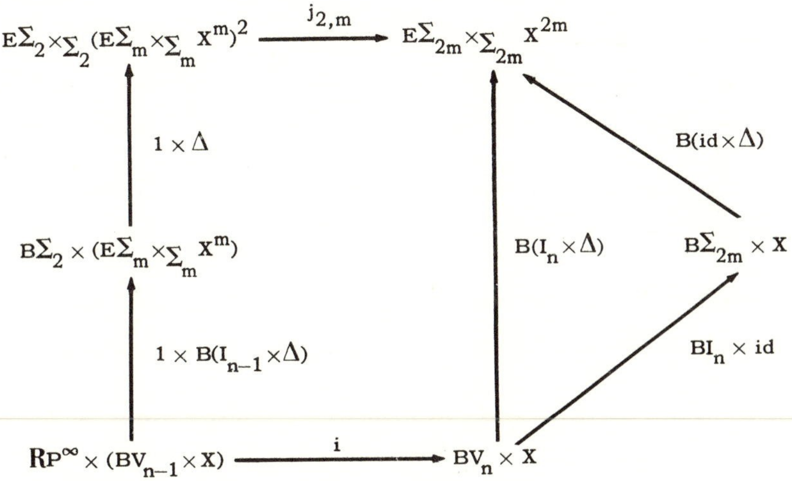

Here $I_n : V_n \to \Sigma_{2^n}$ is the embedding from 3.23 and i arises from the identification $RP^\infty \times BV_{n-1} = BV_n$. Define

3.48 $$St_n(\alpha) = B(I_n \times \Delta)^*(\Gamma_{2^n}(\alpha)) .$$

We have an expansion

$$St_n(\alpha) = \sum_I \theta_I \otimes \alpha^I$$

for some elements $\theta_I \in \mathrm{Image}\,(H^*(B\Sigma_{2^n}) \to H^*(BV_n))$ and elements $\alpha^I \in H^*(X)$ corresponding to them.

LEMMA 3.49. $St_n(\alpha) = \sum_{r=0}^{m} e^r \otimes Sq^{(\dim \alpha)m-r}(St_{n-1}(\alpha))$ *where* e *is the generator in* $H^*(RP^\infty)$ *and* $m = 2^{n-1}$.

Proof. From 3.47 and 3.46 (iii), $St_n(\alpha) = St_1(St_{n-1}(\alpha))$ and the result follows from 3.21.

LEMMA 3.50. *Let* $e \in H^1(RP^\infty)$, *then*

$$St_n(e) = \sum_{i=0}^{n} D_{n-i}/D_n \otimes e^{2^{n-i}}$$

(*in the notation of* 3.24, $D_0/D_n = D_n$).

Proof. First note from the previous lemma that the expansion has the form

$$St_n(e) = \sum \lambda_i \otimes e^{2^i}$$

(since $Sq^I(e) = e^{2^r}$ or 0) with $\lambda_i \in \mathrm{Image}\,(H^*(B\Sigma_{2^n}) \to H^*(BV_n))$. But in our range of dimensions the only image elements are the D_{n-i}/D_n, so we have

$$St_n(e) = \sum \varepsilon_i\, D_{n-i}/D_n \otimes e^{2^{n-i}}, \quad \varepsilon_i \in Z/2 .$$

Let incl: $BV_{n-1} \times X \to RP^\infty \times (BV_{n-1} \times X)$, $X = RP^\infty$. From the previous lemma

$$(\mathrm{incl})^*(\mathrm{St}_n(e)) = (\mathrm{St}_{n-1}(e))^2 ,$$

and inductively we may assume the formula so

$$(\mathrm{incl})^*(\mathrm{St}_n(e)) = \sum (D_{n-i-1}/D_{n-1})^2 \otimes e^{2^{n-i}} .$$

Thus $\varepsilon_i = 1$ except perhaps for ε_n; it remains to check the coefficient of e. But (again from 3.49) this is

$$\sum e^r \otimes Sq^{2^{n-1}-r}(D_{n-1}) = e \otimes D_{n-1}^2 + \cdots \neq 0$$

and the result follows.

COROLLARY 3.51. *Let* $\sigma\colon H^*(X) \to H^{*+1}(S^1 \wedge X)$ *be the suspension isomorphism. Then if* $\mathrm{St}_n(\alpha) = \Sigma\theta_I \otimes \alpha^I$ *we have*

$$\mathrm{St}_n(\sigma(\alpha)) = \sum (D_n \cup \theta_I) \otimes \sigma(\alpha^I) .$$

Proof. We have $\sigma(\alpha) = \bar{e} \cup \alpha$ where $\bar{e}$ is the non-zero class in $H^1(S^1)$. Let $i\colon S^1 \to RP^\infty$ be the non-trivial map then by naturality 3.46 (iv) $\mathrm{St}_n(\bar{e}) = 1 \otimes i^*(\mathrm{St}_n(e)) = D_n \otimes \bar{e}$. Now use 3.46 (i).

Sharpening this result, let $\mathrm{St}^{(i_1,\cdots,i_n)}$ be the element in the Steenrod algebra dual to $\xi_1^{i_1}\xi_2^{i_2}\cdots\xi_n^{i_n}$ in the Milnor monomial basis for $A(2)^*$ ([156]). Then we have

PROPOSITION 3.52. *Let* ι_j *be the fundamental class in* $H^j(K(\mathbf{Z}/2, j))$ *then*

$$\mathrm{St}_n(\iota_j) = \sum w_{2^n-1}^{\varepsilon} \cdot w_{2^n-2}^{i_1} \cdots w_{2^n-1}^{i_{n-1}} \otimes \mathrm{St}^{(i_1,\cdots,i_n)}(\iota_j)$$

where $i_1 + \cdots + i_n + \varepsilon = j$ *and* $w_{2^n-2^k} = D_k/D_n$.

Proof. Note that $H^*(K(\mathbf{Z}/2, j))$ is a polynomial algebra in certain $\mathrm{St}^I(\iota_j)$ where St^I runs over all operations of excess $< j$ (i.e. $i_1 + \cdots + i_{n-1} < j$). From 3.49 we must have

$$\mathrm{St}_n(\iota_j) = \sum \theta_I \otimes \mathrm{St}^I(\iota_j)$$

for some polynomial θ_I in the classes $w_{2^n-1}, \cdots, w_{2^n-1}$ since the St^I form a basis of the Steenrod algebra. Now, use the formula

$$\Gamma_{2^n}(e_1 \otimes \cdots \otimes e_j) = \Gamma_{2^n}(e_1) \cup \cdots \cup \Gamma_{2^n}(e_j)\,.$$

This gives, in $(RP^\infty)^j$

$$\mathrm{St}_n(e_1 \otimes \cdots \otimes e_j) = \sum w^{\varepsilon}_{2^n-1} w^{i_1}_{2^n-2} \cdots w^{i_{n-1}}_{2^{n-1}} \otimes \mathrm{St}^{(i_1, \cdots, i_n)}(e_1 \otimes \cdots \otimes e_j)\,.$$

Indeed, the formula is true for $j = 1$ by 3.50 and it follows in general using $\mathrm{St}_n(e_1 \otimes \cdots \otimes e_j) = \mathrm{St}_n(e_1 \otimes \cdots \otimes e_{j-1}) \cup \mathrm{St}_n(e_j)$ and the Cartan formula for St^I:

$$\mathrm{St}^I(x \cup y) = \sum_{I'+I''=I} \mathrm{St}^{I'}(x) \cup \mathrm{St}^{I''}(y)\,.$$

In dimensions less than $2t$ the map

$$e_1 \otimes \cdots \otimes e_t : (RP^\infty)^t \to K(\mathbf{Z}/2, t)$$

induces an injection in cohomology, so by naturality and the suspension result 3.51 the formula follows.

We can now use the slant product $H_*(X) \otimes H^*(X \times Y) \to H^*(Y)$ to define a map

$$C_r : H_i(B\Sigma_r) \to H^{rj-i}(K(\mathbf{Z}/2, j))\,.$$

Let $\bar{\Delta}: B\Sigma_r \times K(\mathbf{Z}/2, j) \to E\Sigma_r \times_{\Sigma_r} K(\mathbf{Z}/2, j)^r$ be the obvious diagonal mapping. We define

$$C_r(x) = x/\bar{\Delta}^*(\Gamma_r(\iota_j))\,.$$

The calculations above show

$$C_{2^n}(E_{(\varepsilon, i_1, \cdots, i_{n-1})}) = \mathrm{St}^{(i_1, \cdots, i_n)}(\iota_j)\,,$$

where $\varepsilon + i_1 + \cdots + \iota_{n-1} + \iota_n = j$, and if $\varepsilon > 0$ then $\mathrm{Excess}(\mathrm{St}^{(i_1, \cdots, i_n)}) = i_1 + \cdots + i_n < j$ so $\mathrm{St}^{(i_1, \cdots, i_n)}(\iota_j)$ is a polynomial generator of $H^*(K(\mathbf{Z}/2, j))$.

From 4.46 (ii) we get the commutative diagram

$$\begin{array}{ccc}
H_i(B\Sigma_r) \otimes H_\ell(B\Sigma_s) & \xrightarrow{\quad * \quad} & H_{i+\ell}(B\Sigma_{r+s}) \\
\downarrow & & \downarrow \\
H^{rj-i}(K(\mathbf{Z}/2, j)) \otimes H^{sj-\ell}(K(\mathbf{Z}/2, j)) & \xrightarrow{\text{cup}} & H^{(r+s)j-(i+\ell)}(K(\mathbf{Z}/2, j))
\end{array}$$

where $*$ is induced from the usual inclusion $\Sigma_r \times \Sigma_s \subset \Sigma_{r+s}$. The inclusion $X \to E\Sigma_r \times_{\Sigma_r} X^r$ maps $\Gamma_r(a)$ to a^r (by definition) (cf. 3.49) so the diagram above specializes to show commutativity in

$$\begin{array}{ccc}
H_i(B\Sigma_r) & \xrightarrow{\quad *[s] \quad} & H_i(B\Sigma_{r+s}) \\
\big\downarrow C_r & & \big\downarrow C_{r+s} \\
H^{jr-i}(K(\mathbf{Z}/2, j)) & \xrightarrow{\ \cup\, \iota_j^s\ } & H^{j(r+s)-i}(K(\mathbf{Z}/2, j))
\end{array}$$

where $[s] \in H_0(B\Sigma_s)$ is the non-zero element, and $*[s]$ the map induced from the inclusion $\Sigma_r \subset \Sigma_{r+s}$.

Using these calculations we can now finish the proof of Theorem 3.32 directly. From 3.23, 3.28, 3.30 we have that $H_*(C(S^0))$ is generated by the elements

$$E_{(\varepsilon, i_1, \cdots, i_{n-1})} * [s] \in H_*(B\Sigma_{2^n+s}) \subset H_*(C(S^0)), \ \varepsilon > 0 .$$

Now C_{2^n+s} maps $E_{(\varepsilon, i_1, \cdots, i_{n-1})} * [s]$ to $\mathrm{St}^{(i_1, \cdots, i_n)}(\iota_j) \cup \iota_j^s$ in $H^*(K(\mathbf{Z}/2, j))$ and if we pick j larger than the dimensions under consideration then $i_n \neq 0$, so the corresponding elements $\mathrm{St}^{(i_1, \cdots, i_n)}(\iota_j)$ are distinct polynomial generators of $H^*(K(\mathbf{Z}/2, j))$.

This completes the proof.

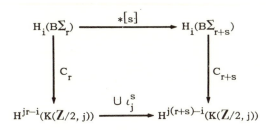

CHAPTER 4
THE HOMOTOPY STRUCTURE OF G/PL AND G/TOP

In this chapter we review the work due to Sullivan, which describes the homotopy types of G/PL and G/TOP. In Chapter 2 we saw that these spaces classify "surgery problems" (degree 1 normal maps) in the PL and topological categories, respectively. In particular, if M^n is a closed smooth manifold, then the simply connected surgery obstructions define maps

4.1 $$s : [M^n, G/PL] \to P_n ,$$

where $P_{4k} = \mathbb{Z}$, $P_{4k+2} = \mathbb{Z}/2$ and $P_{2k+1} = 0$. Indeed, let

$$f : \hat{M}^n \to M^n , \qquad \hat{f} : \nu(\hat{M}) \to \xi$$

denote the surgery problem associated to $\gamma : M^n \to G/PL$. If $n = 4k$, and M is oriented then

$$s(M, \gamma) = \frac{1}{8} (\text{Index } \hat{M} - \text{Index } M) ,$$

and if $n = 4k+2$, then $s(M, \gamma)$ is the Kervaire invariant of $(f, \hat{f})$.

The obstruction for $n = 4k$, is a difference of cobordism invariants, and thus factors through a similar map

4.2 $$s_I : \Omega_{4k}(G/PL) \to \mathbb{Z} .$$

It is less evident that the Kervaire invariant is a cobordism invariant, but this is also true. For a relatively elementary geometric proof see [117]. A more general proof is obtained from the homotopy theoretical definition of the Kervaire invariant [27]. Thus the Kervaire invariant factors through a map

4.3 $$s_K : \mathfrak{N}_{4k+2}(G/PL) \to \mathbb{Z}/2 .$$

Sullivan's basic idea in studying G/PL and G/TOP was to use the connection between bordism and ordinary homology or K-homology to reinterpret 4.2 and 4.3 in terms of characteristic classes in cohomology or K-theory, thus getting maps of G/PL into Eilenberg-Maclane spaces or BO which give homotopy equivalences on localizing. (Low dimensional calculations at odd primes were initially done by F. Peterson.)

A. *The 2-local homotopy type of* G/PL

The unoriented Thom spectrum MO and the 2-local oriented Thom spectrum MSO[2] are both wedges of Eilenberg-Maclane spectra by 1.27 and 1.35. Hence there exist sections $T_2 : K(Z/2) \to MO$ or $T : K(Z_{(2)}) \to MSO[2]$ of the Thom classes $U_2 : MO \to K(Z_{(2)})$ or $U : MSO[2] \to K(Z_{(2)})$*) giving natural homomorphisms

4.4 $$t_2 : H_*(X; Z/2) \to \mathfrak{N}_*(X), \quad t : H_*(X; Z_{(2)}) \to \Omega_*(X) \otimes Z_{(2)} .$$

The compositions

$$S^k \xrightarrow{\iota_k} K(Z/2, k) \xrightarrow{T_2} MO_k$$

$$S^k \xrightarrow{\iota_k} K(Z_{(2)}, k) \xrightarrow{T} MSO_k[2]$$

are homotopic to the inclusions of a fiber. It follows that the compositions of t_2, t with the ordinary Hurewicz map

4.5 $$\pi_n(X) \xrightarrow{h} H_n(X; Z/2) \xrightarrow{t_2} \mathfrak{N}_n(X)$$
$$\pi_n(X) \xrightarrow{h} H_n(X; Z_{(2)}) \xrightarrow{t} \Omega_n(X) \otimes Z_{(2)}$$

are the cobordism Hurewicz maps, which to a homotopy class of maps $f: S^n \to X$ associate the cobordism classes $\{S^n, f\}_2 \in \mathfrak{N}_n(X)$, $\{S^n, f\} \in \Omega_n(X)$, respectively

*) Canonical sections were constructed by Mahowald, see [155].

For $X = G/PL$ we can compose the maps from 4.4 with the maps s_K, s_I to get homomorphisms

4.6
$$K_{4n-2} \in \mathrm{Hom}(H_{4n-2}(G/PL; Z/2), Z/2)$$
$$K_{4n} \in \mathrm{Hom}(H_{4n}(G/PL; Z_{(2)}), Z_{(2)})$$

or equivalently cohomology classes $K_{4n-2} \in H^{4n-2}(G/PL; Z/2)$, $K_{4n} \in F^{4n}(G/PL) \otimes Z_{(2)}$ (see p. 24 for the definition of $F^*(X)$).

The natural map

$$H^*(G/PL; Z_{(2)}) \to F^*(G/PL) \otimes Z_{(2)}$$

is surjective and we can choose liftings of the classes K_{4n} to classes $\overline{K}_{4n} \in H^{4n}(G/PL; Z_{(2)})$.

LEMMA 4.7. *The classes* $\overline{K}_{4n}$ *and* $K_{4n-2} (n > 1)$ *define a map*

$$\phi : G/PL[2] \to \prod_{n>1} K(Z_{(2)}, 4n) \times K(Z/2, 4n-2)$$

whose homotopy fiber $F(\phi)$ *is a two stage Postnikov system with non-zero homotopy groups only* $\pi_2(F(\phi)) = Z/2$ *and* $\pi_4(F(\phi)) = Z_{(2)}$.

Proof. It suffices to show that ϕ induces isomorphisms on homotopy groups in dimensions larger than 4, since from 2.24

$$\pi_n(G/PL[2]) = \begin{cases} Z_{(2)} & \text{for } n \equiv 0 \pmod 4 \\ Z/2 & \text{for } n \equiv 2 \pmod 4 \\ 0 & \text{for } n \equiv 1 \pmod 2 . \end{cases}$$

The generators $\iota_{2n} \in \pi_{2n}(G/PL[2])$ are specified by

$$s_I(S^{4n}, \iota_{4n}) = 1 \text{ for } n > 1$$
$$s_K(S^{4n-2}, \iota_{4n-2}) = 1 \text{ for } n \geq 1$$

so K_{4n}, K_{4n-2} evaluate to 1 on the homotopy generators in dimension $\neq 4$ (cf. 4.5), and 4.7 follows.

THEOREM 4.8. *The 2-local homotopy type of* G/PL *is given by*

$$G/PL[2] \simeq E \times \prod_{n>1} K(Z_{(2)}, 4n) \times K(Z/2, 4n-2) ,$$

where E *is the fiber in the fibration*

$$E \longrightarrow K(Z/2, 2) \xrightarrow{\beta Sq^2} K(Z_{(2)}, 5)$$

and β *denotes the Bockstein operator.*

Proof. It suffices to check that the fiber $F(\phi)$ in 4.7 has K-invariant βSq^2 since it is evident that

$$G/PL[2] \simeq F(\phi) \times \prod_{n>1} K(Z_{(2)}, 4n) \times K(Z/2, 4n-2) .$$

Since PL/O is 6-connected [39], G/PL $\simeq$ G/O through at least dimension 5. Thus it suffices to analyze the fiber G/O of BO $\to$ BG in low dimensions which in view of Chapter 3 and Bott's results is direct:*) One checks that $\pi_4(G/O) \xrightarrow{h} H_4(G/O; Z_{(2)})$ is not an isomorphism, so $F(\phi)$ must have a non-zero K-invariant. Since G/PL is deloopable the K-invariant must be stable and βSq^2 is then the only possibility.

There is a great deal of ambiguity in the classes K_{4n-2}, K_{4n} in 4.6, and in the choice of $\bar{K}_{4n}$ lifting K_{4n} in 4.7, and consequently the map ϕ is not unique. We now indicate how to get unique classes K_{4n-2}, K_{4n} in 4.6. (Unique liftings $\bar{K}_{4n}$ are more delicate and are obtained only after a great deal of effort in [91] and [103], cf. the footnote to 4.32.)

*)In the next chapter we show that $G/O[2] \simeq BSO[2]$ through dimension 5. This gives an alternative argument.

Let $\varepsilon: \Omega_*(\ ) \to H_*(\ ;\mathbf{Z})$ be the augmentation map (induced from the Thom class $U: MSO \to K(\mathbf{Z})$, see 1.40). It induces

$$\varepsilon^*: \mathrm{Hom}(H_*(G/PL;\mathbf{Z});\mathbf{Z}_{(2)}) \to \mathrm{Hom}(\Omega_*(G/PL);\mathbf{Z}_{(2)})$$

but s_I is not in its image. Indeed $\mathrm{Im}(\varepsilon^*)$ vanishes on decomposables (1.40) but the surgery problem over a composite $M \times N \xrightarrow{p_2} N \xrightarrow{\gamma} G/PL$ is

$$\mathrm{Id} \times f: M \times \hat{N} \to M \times N,\ \mathrm{Id} \times \hat{f}: \nu(M) \times \nu(\hat{N}) \to \nu(M) \times \xi\,,$$

so

$$s_I(M \times N, \gamma p_2) = s_I(N, \gamma) \cdot \mathrm{Index}(M)\,.$$

Similarly, Sullivan proved a product formula for the Kervaire invariant

$$s_K(M \times N, \gamma p_2) = s_K(N;\gamma)\chi(M)$$

if $\dim N = 4k+2$, where $\chi(M)$ is the mod. 2 Euler characteristic: $\chi(M) = [\Sigma \dim H^i(M;\mathbf{Z}/2)](\mathrm{mod.}\ 2)$. (For a proof of this product formula see [27] or [117].)

THEOREM 4.9. *There exist unique classes* $K_{4n} \in F^{4n}(G/PL;\mathbf{Z}_{(2)})$ *and* $K_{4n-2} \in H^{4n-2}(G/PL;\mathbf{Z}/2)$ *such that*

$$s_I(M,\gamma) = \langle \mathcal{L}(\tau(M)) \cdot \sum_{n \geq 1} \gamma^*(K_{4n}), [M] \rangle$$

$$s_K(M,\gamma) = \langle V^2(\tau(M)) \sum_{n \geq 1} \gamma^*(K_{4n-2}), [M] \rangle$$

where $\mathcal{L}$ *is the Hirzebruch genus from* 1.39 *and* $V^2(\xi)$ *is the square of the total Wu class.**) *Moreover*

$$\psi(K_{4n}) = K_{4n} \otimes 1 + 1 \otimes K_{4n} + 8 \quad K_{4i} \otimes K_{4n-4i},\ \psi(K_{4n-2}) = K_{4n-2} \otimes 1 + 1 \otimes K_{4n-2}\,.$$

---

*)The total Wu class $V(\xi)$ is related to the total Stiefel-Whitney class by the formula $SQ(V(\xi)) = W(\xi)$, i.e. $w_m(\xi) = \Sigma Sq^i(v_{m-i}(\xi))$. For $\xi = \tau(M)$, the tangent bundle of a 2n-dimensional manifold, $v_i(\tau(M)) = 0$ for $i > n$ and $v_n(\tau(M))^2 = w_{2n}(\tau M)$. Hence $\chi(M) = \langle v_n(\tau(M)^2, [M] \rangle$.

Proof. In the index case define K_{4i} inductively by first defining a new homomorphism

$$\tilde{K}_{4i} : \Omega_{4i}(G/PL) \to Z_{(2)}$$

by setting

$$\tilde{K}_{4i}(M^{4i}, \gamma) = s_I(M, \gamma) - \langle \sum_{j<i} \gamma^*(K_{4j}) L_{4(i-j)}(\tau(M)), [M] \rangle . \tag{4.10}$$

The only thing to observe is that $L_{4i}(\tau(M)) \in H^*(M; Z_{(2)})$. But this follows from 1.35(a) and 1.39 which together show

$$\mathcal{L}^{-1} \in H^*(BO; Z_{(2)}) ,$$

and the fact that $\mathcal{L} = \chi(\mathcal{L}^{-1})$.

Note the following calculation

$$\tilde{K}_{4i}(N \times M, \gamma p_2) = \text{Index}(N)\, s_I(M, \gamma) - \langle \mathcal{L}(\tau(N)), [N] \rangle \langle \sum_{j \geq 1} \gamma^*(K_{4j}) \mathcal{L}(\tau(M)), [M] \rangle = 0$$

by the inductive assumption. Then $\tilde{K}_{4i}$ factors through

$$\Omega_*(G/PL) \otimes_{\Omega_*(pt)} Z_{(2)} \xrightarrow{\cong} H_*(G/PL; Z_{(2)})$$

(cf. 1.40) and it defines an element $K_{4i} \in F^{4i}(G/PL; Z_{(2)})$. Then using 4.10 this gives the first part of 4.9.

To obtain the second part we work in a similar way using $\mathfrak{N}_*(G/PL)$ and the identification

$$\mathfrak{N}_*(G/PL) \otimes_{\mathfrak{N}_*(pt)} Z/2 \xrightarrow{\cong} H_*(G/PL; Z/2) .$$

The key observation is that $\langle V^2(\tau(M)), [M] \rangle = \chi(M)$.

B. *Ring spectra, orientations and* K-*theory at odd primes*

The analysis in Chapter 4A was possible because of 1.27, 1.35. The analysis at odd primes is similar in spirit but differs in details since away from $2, \Omega_*(\ )$ is more directly connected with K-theory than homology. Now we review briefly these connections.

A *ring spectrum* $E = \{E_k, f_k\}$ is a spectrum with the following additional structure,

4.11
$$\mu_{k,\ell}: E_k \wedge E_\ell \to E_{k+\ell} \quad \text{(multiplication)}$$
$$\iota_k: \; S^k \to E^k \qquad \text{(unit)}$$

such that $\mu_{k,\ell}\circ(1\wedge\iota_\ell)$ and $\mu_{k,\ell}\circ(\iota_k\wedge 1)$ are related to the structure maps for E in the obvious fashion ([3]).

The Thom spectra MSO, MSPL and MSTOP are ring spectra. The multiplication in each case is induced from Whitney sum of the universal bundles over BSO_k, $BSPL_k$ and $BSTOP_k$.

If Λ is a ring, then the Eilenberg-MacLane spectrum $K(\Lambda)=\{K(\Lambda,k)\}$ becomes a ring spectrum when we set

$$\mu^*_{k,\ell}(\bar{\iota}_{k+\ell}) = \bar{\iota}_k \otimes \bar{\iota}_\ell \,.$$

($\bar{\iota}_k$ is the fundamental class of $H^k(K(\Lambda,k);\Lambda)$.)

The spectrum for periodic KO-theory is a ring spectrum. Its 8k'th space is a copy of $BO\times Z$ and the multiplication $(BO\times Z)\wedge(BO\times Z)\to BO\times Z$ is the map induced from tensor product of stable vector bundles.

A basic example for our purpose is the KO-spectrum above localized away from 2, whose 8k'th space is $BO[\frac{1}{2}]\times Z[\frac{1}{2}]$. From 1.33 we have $\Omega^4(BO[\frac{1}{2}]\times Z[\frac{1}{2}]) = BO[\frac{1}{2}]\times Z[\frac{1}{2}]$.

A map of spectra $\phi: F\to E$ consists of a sequence of maps $\phi_k: F_k\to E_k$ satisfying the obvious compatibility conditions. Let E be any ring spectrum. An E-*orientation* of F is a map

$$\Delta: F \to E$$

such that $\Delta\circ\iota_F \simeq \iota_E$ where ι_F and ι_E are the units.

Specifying the Thom class U generating $H^0(MSO; Z)$ is equivalent to a map of spectra

4.12
$$u: MSO \to K(Z)$$

so $u^*(\iota) = U$, which is directly seen to be an orientation. Similarly, MSPL and MSTOP are $K(Z)$-oriented.

The E-orientation is said to be *multiplicative* if Δ preserves the ring structure, that is the Δ_k are compatible with the $\mu_{k,\ell}$. (The orientation in 4.12 is multiplicative.)

THEOREM 4.13 (Dold [45]). *Let* $\Delta : MSO \to E$ *be an orientation, and suppose* ξ *is an oriented* k*-dimensional vector bundle over* X. *Set* $\Delta(\xi) \in \tilde{E}^k(M(\xi))$ *to be the composite.*

$$M(\xi) \longrightarrow MSO_k \xrightarrow{\Delta_k} E_k .$$

Then

$$\Delta(\xi) \cup : E^i(X) \to \tilde{E}^{i+k}(M(\xi))$$

is an isomorphism.

(The proof is analogous to that of the usual Thom isomorphism. Indeed it is clear if ξ is a trivial bundle and it follows in general from a spectral sequence argument or equivalently, a piecing together argument using the Meyer-Vietoris sequence.)

We now turn to K-theory. Let $KO^*(\ ; Z[\frac{1}{2}])$ and $KO_*(\ ; Z[\frac{1}{2}])$ denote the $Z/4$-graded orthogonal K-cohomology and K-homology theories (based on the 4-fold periodic spectrum $BO[\frac{1}{2}] \times Z[\frac{1}{2}]$).

The coefficient groups $\tilde{KO}^0(S^n; Z[\frac{1}{2}])$ are zero for $n \not\equiv 0 \pmod 4$ and a single copy of $Z[\frac{1}{2}]$ for $n \equiv 0 \pmod 4$. We fix a generator $a \in \tilde{KO}^0(S^4; Z[\frac{1}{2}])$ to be the element which under complexification

$$c : \tilde{KO}^0(S^4; Z[\tfrac{1}{2}]) \to \tilde{K}^0(S^4; Z[\tfrac{1}{2}])$$

maps to $\frac{1}{2}b^2$, where $b \in \tilde{K}^0(S^2)$ is represented by the reduced canonical complex line bundle, $b = H - 1$. The powers of a generate the other non-zero coefficient groups,

$$\tilde{KO}^0(S^{4n}; Z[\tfrac{1}{2}]) = Z[\tfrac{1}{2}](a^n) .$$

The spectrum for $\widetilde{KO}^*(\ ;Z[½])$ has a copy of $BO[½] \times Z[½]$ in each degree $4k$ and the structure maps

$$S^4 \wedge (BO[½] \times Z[½]) \to BO[½] \times Z[½]$$

are induced by multiplication with a.

The Chern character

$$ch : K^0(X) \to H^*(X; Q)$$

is the unique ring homomorphism whose value on a line bundle H is given by $ch(H) = e^{c_1(H)}$ where c_1 is the first Chern class (cf. [54]). Clearly ch maps $K^0(S^2)$ isomorphically onto $H^*(S^2; Z)$ and hence in general maps $K^0(S^{2n})$ isomorphically onto $H^*(S^{2n}; Z)$. Composing with complexification we get the Pontrjagin character

$$ph : KO^0(X) \to H^*(X; Q)\ ,$$

$ph = ch \circ c$. We note that $ph(a^n)$ is the generator of $H^{4n}(S^{4n}; Z)$ where $a \in KO^0(S^4; Z[½])$ is the generator defined above.

From the identification in 1.33, $MSpin[½] \simeq MSO[½]$, the BO orientation of MSpin induces a $BO[½]$ orientation of $MSO[½]$. For our purpose, however, it is more convenient to use the $BO[½]$ orientation from ([133], Chapter IX)

$$\Delta : MSO \to BO[½]$$

whose Pontrjagin character is given as

4.14 $$ph\,\Delta = \mathcal{L}^{-1} \cup U$$

in $H^*(MSO; Q)$. Then the induced natural transformation

$$\delta : \Omega_*(X) \to KO_*(X; Z[½])$$

in degrees $4k$ and for $X = pt$ reduces to the index homomorphism,

$$\delta(\{M^{4k}\}) = \text{Index}\,(M^{4k})\cdot a^n\,.$$

Thus, if we consider $Z[\frac{1}{2}]$ as an $\Omega_*(pt)$-module via the index homomorphism δ factors to define a homomorphism.

4.15 $$\bar{\delta}: \Omega_{4*+i}(X)\otimes_{\Omega_*(pt)} Z[\tfrac{1}{2}] \xrightarrow{\cong} KO_i(X; Z[\tfrac{1}{2}])$$

$(i \in Z/4)$ and by results of Conner and Floyd [42] $\bar{\delta}$ is an isomorphism.

Our use of 4.15 in this and the next chapter will be to factor certain homomorphisms of $\Omega_*(X)$ through homomorphisms of $KO_*(X)$. We will then need a universal coefficient theorem to reinterpret such homomorphisms as elements of $KO^*(X)$. First, we recall the Kronecker pairing

$$<,>: KO^0(X; Z[\tfrac{1}{2}]) \otimes KO_0(X; Z[\tfrac{1}{2}]) \to Z[\tfrac{1}{2}]\,.$$

Suppose given elements $\xi \in KO^0(X; Z[\frac{1}{2}])$ and $x \in KO_0(X; Z[\frac{1}{2}])$ represented by based maps

$$\xi: X^+ \to BO[\tfrac{1}{2}] \times Z[\tfrac{1}{2}]$$

$$x: S^{4n} \to X^+ \wedge (BO[\tfrac{1}{2}] \times Z[\tfrac{1}{2}])\,.$$

Then $<\xi, x> \in Z[\frac{1}{2}]$ is the number such that $<\xi, x>\cdot a^n$ is represented by the composite

$$S^{4n} \xrightarrow{x} X^+ \wedge (BO[\tfrac{1}{2}]\times Z[\tfrac{1}{2}]) \xrightarrow{\xi\wedge 1}$$

$$(BO[\tfrac{1}{2}]\times Z[\tfrac{1}{2}]) \wedge (BO[\tfrac{1}{2}]\times Z[\tfrac{1}{2}]) \xrightarrow{\otimes} BO[\tfrac{1}{2}] \times Z[\tfrac{1}{2}]\,.$$

THEOREM 4.16 (D. Anderson). *The homomorphism*

$$\text{eval}: KO^0(X; Z[\tfrac{1}{2}]) \to \text{Hom}(KO_0(X; Z[\tfrac{1}{2}]), Z[\tfrac{1}{2}])$$

adjoint to the Kronecker pairing is an epimorphism with kernel $\text{Ext}\,(KO_0(SX); Z[\frac{1}{2}])$ *when* X *is a finite* CW *complex.*

(For a proof see [147].) It is worth noting that if $H_*(X; Z[\frac{1}{2}])$ is concentrated in even degrees then eval is an isomorphism, since in this case the Atiyah-Hirzebruch spectral sequence

$$H_*(X; KO_*(pt; Z[\tfrac{1}{2}])) \Longrightarrow KO_*(X; Z[\tfrac{1}{2}])$$

implies that

$$KO_0(SX; Z[\tfrac{1}{2}]) = 0 .$$

C. *Piece-wise linear Pontrjagin classes*

In Chapter 1 we defined the Pontrjagin classes for vector bundles and then showed that certain complicated polynomials (the $\mathfrak{L}$-genus) in them were related to the index of 4k-dimensional smooth manifolds. The coefficient of p_{4i} in L_{4i} is non-zero, so rationally we can solve to express the Pontrjagin classes as polynomials in the classes L_{4i}. But PL manifolds M^{4k} also have an index and we can directly construct the classes L_{4i} in the PL-case in analogy with the construction of the universal surgery classes in Chapter 4A. (Cf. Thom [138].)

We consider the homomorphism

4.17 $$I: \Omega_{k+i}(MSPL_k) \to Z \qquad (k >> i)$$

which associates to a singular manifold (V^{k+i}, f) in $MSPL_k$ the index of the PL manifold $M^i = f^{-1}(BSPL_k)$, where we have assumed f to be transverse to $BSPL_k \subset MSPL_k$. (This is well defined by an easy transversality argument [146].) If $i \not\equiv 0 \pmod 4$ we set $I\{V, f\} = 0$.

THEOREM 4.18. *There is a unique multiplicative class*

$$\mathfrak{L}_{PL}^{-1} = 1 + \bar{L}_4 + \bar{L}_8 + \cdots + \bar{L}_{4n} + \cdots$$

in $F^*(BSPL; Z_{(2)})$ *such that*

$$(*) \qquad I(\{V,f\}) = \langle f^*(\mathcal{L}_{PL}^{-1} \cdot U_{PL}) \cdot \mathcal{L}\{V\}, [V] \rangle$$

where I *is the homomorphism in* 4.17, $\mathcal{L}\{V\}$ *is defined in* 1.39 *and* $U_{PL} \in H^0(MSPL; Z)$ *is the Thom class.*

Proof. The proof is similar to 4.9. Suppose inductively that $\bar{L}_4, \cdots, \bar{L}_{4i-4}$ are defined, that $\psi(\bar{L}_{4r}) = \sum \bar{L}_{4j} \otimes \bar{L}_{4r-4j}$ for $r < i$ and that (*) is satisfied when $\dim V < k+4i$.

Consider

$$\tilde{I}\{V^{k+4i}, f\} = I\{V^{k+4i}, f\} - \sum_{j=0}^{i-1} \langle f^*(\bar{L}_{4j} \cdot U_{PL}) \cdot L_{4(i-j)}(V), [V] \rangle$$

where $L_{4(i-j)}(V)$ is the $4(i-j)$ dimensional component of the Hirzebruch genus of $\tau(V)$. If $\{V, f\}$ is "decomposable,"

$$\{V,f\} = \{W^{4w}, g\} \cdot \{N^{4n}\}, \qquad (4n > 0)\ .$$

and we choose f carefully ($f = g \circ \text{proj}$), then

$$f^{-1}(BSPL_k) = g^{-1}(BSPL_k) \times N\ .$$

On the other hand,

$$\sum_{j=0}^{i-1} \langle f^*(\bar{L}_{4j} U_{PL}) \cdot L_{4(i-j)}(W \times N), [W \times N] \rangle =$$

$$\sum \langle g^*(\bar{L}_{4j} \cdot U_{PL}) L_{4\nu}(W) \otimes L_{4\mu}(N), [W] \otimes [N] \rangle$$

$$= \sum_{j=0}^{w} \langle g^*(\bar{L}_{4j} \cdot U_{PL}) \cdot L_{4(w-j)}(W), [W] \rangle \cdot \langle L_{4n}(N), [N] \rangle = I\{W, g\} \cdot \text{Index}\{N\}$$

where the last equation follows from the inductive hypothesis and 1.39. Thus $\tilde{I}(\{W, g\} \cdot \{N\}) = 0$ and $\tilde{I}$ induces a homomorphism

$$\tilde{I}: H_{k+4i}(MSPL_k; Z_{(2)}) \to Z_{(2)}$$

hence a class $\tilde{I}_{4i} \in F^{k+4i}(MSPL_k; Z_{(2)})$ which we can write as $I_{4i} = \bar{L}_{4i} \cup U_{PL}$ for a unique class $\bar{L}_{4i} \in F^{4i}(BSPL_k; Z_{(2)})$. This completes the inductive step.

Let $\mathcal{L}_{PL}$ be the "inverse" of the class $\mathcal{L}_{PL}^{-1}$ from 4.18, that is, $\mathcal{L}_{PL} = \chi(\mathcal{L}_{PL}^{-1})$, where $\chi: F^*(BSPL; Z_{(2)}) \to F^*(BSPL; Z_{(2)})$ is induced from the map $\chi: BSPL \to BSPL$ which classifies $-\xi$. Then $\mathcal{L}_{PL}^{-1}(\nu(M)) = \mathcal{L}_{PL}(\tau(M))$ and we have as a special case of 4.18 the PL version of the index theorem,

COROLLARY 4.19. *For a* 4k-*dimensional* PL-*manifold* M,

$$\text{Index}(M) = \langle \mathcal{L}_{PL}(\tau(M)), [M] \rangle .$$

DEFINITION 4.20. Let $\mathcal{L}(\sigma_4, \sigma_8, \cdots)$ be the genus in 1.39 and define $p_{4i} \in H^{4i}(BSPL; Q)$ to be the unique classes such that

$$\mathcal{L}_{PL} = \mathcal{L}(p_4, p_8, \cdots, p_{4i}, \cdots)$$

in $H^*(BSPL; Q)$.

It is direct that $p_{4i} \in H^{4i}(BSPL; Q)$ restricts to the rational reduction of the usual (integral) Pontrjagin class under the inclusion $BSO \to BSPL$. The PL Pontrjagin classes, however, are not in general integral, not even $Z_{(2)}$-integral ($p_{4i} \in H^{4i}(BSPL; Q)$ comes from a $Z_{(2)}$-integral class for small values of i, $i < 30$, but e.g. $p_{120} \in H^{120}(BSPL; Q)$ is not the rational reduction of a $Z_{(2)}$-integral class. This is examined in more details in Chapters 10 and 11).

REMARK 4.21. The fiber PL/O of the natural map $BSO \to BSPL$ has finite homotopy groups: $\pi_n(PL/O)$ can be identified with the group of smoothings of the n-sphere for $n \geq 5$ and this group is finite (Kervaire-Milnor [61]). For $n \leq 4$ the finiteness follows from Cerf [39]; in fact

$\pi_n(PL/O) = 0$ for $n \leq 6$. Thus the natural map induces an isomorphism $H^*(BSPL; Q) \xrightarrow{\simeq} H^*(BSO; Q)$, giving an alternative (and less elementary) approach to the rational PL Pontrjagin classes.

The natural map $\hat{j}: G/PL \to BSPL$ is a rational equivalence since its fiber SG has finite homotopy groups by 3.8 and Serre's result from [122]. Our next result gives the connection between the total class $K = K_4 + K_8 + \cdots \in F^*(G/PL; Q)$ and the class $\mathcal{L}_{PL} \in F^*(BSPL; Q)$.

COROLLARY 4.22. $\hat{j}^*(\mathcal{L}_{PL}) = 1 + 8K$.

Proof. The degree 1 normal map associated with $\gamma: M \to G/PL$ has the form

$$f: \hat{M} \to M, \hat{f}: \nu(\hat{M}) \to \zeta$$

with $\zeta = \nu(M) - \xi$ and where ξ is the PL bundle associated to $j \circ \gamma: M \to BSPL$. Now, using 4.19

$$\begin{aligned} \text{Index}(\hat{M}) &= \langle \mathcal{L}_{PL}(\tau(\hat{M})), [\hat{M}] \rangle \\ &= \langle \mathcal{L}_{PL}(\tau(M) \oplus \xi), [M] \rangle \\ &= \langle \mathcal{L}_{PL}(\tau(M)) \cdot \mathcal{L}_{PL}(\xi), [M] \rangle . \end{aligned}$$

Hence

$$\text{Index } \hat{M} - \text{Index } M = \langle \mathcal{L}_{PL}(\tau(M))(\mathcal{L}_{PL}(\xi) - 1), [M] \rangle$$

and we compare with 4.9 to get $\mathcal{L}_{PL}(\xi) - 1 = \gamma^*(8K)$ as claimed.

D. *The homotopy type of* $G/PL[\frac{1}{2}]$

The homomorphism $s_I: \Omega_{4*}(G/PL) \to Z[\frac{1}{2}]$ from 4.2 factors over $\Omega_{4*}(G/PL) \otimes_{\Omega_*(pt)} Z[\frac{1}{2}]$ where $\Omega_*(pt)$ acts on $Z[\frac{1}{2}]$ via the index homomorphism. This follows from the product formula

$$s_I(M \times N, \gamma p_2) = s_I(N, \gamma) \cdot \text{Index}(M) .$$

Hence we get a homomorphism (cf. 4.15)

$$\sigma_0 : KO_0(G/PL; Z[½]) \to Z[½] .$$

In the proof of the next lemma and at various points later in the book we shall use the inverse limit functor $(\varprojlim)$ and its derived functor $(\varprojlim^{(1)})$, so we recall its definition (for especially simple indexing sets). Let

$$A_1 \xleftarrow{f_1} A_2 \xleftarrow{f_2} A_3 \longleftarrow \cdots$$

be a system of abelian groups A_i and homomorphisms f_i. Define

$$F : \prod_{i=1}^{\infty} A_i \to \prod_{i=1}^{\infty} A_i$$

to be the homomorphism

$$F(a_1, a_2, a_3, \cdots) = (a_1 - f_1(a_2), a_2 - f_2(a_3), \cdots)$$

and set

$$\varprojlim A_i = \operatorname{Ker} F , \qquad \varprojlim^{(1)} A_i = \operatorname{coker} F .$$

Suppose we are given a short exact sequence of directed abelian groups as above

$$0 \longrightarrow \{A_i, f_i\} \xrightarrow{\{\alpha_i\}} \{B_i, g_i\} \xrightarrow{\{\beta_i\}} \{C_i, h_i\} \longrightarrow 0 ,$$

that is, $\alpha_i \circ f_i = g_i \circ \alpha_{i+1}$ and similarly for β_i, and for each i the sequence

$$0 \longrightarrow A_i \xrightarrow{\alpha_i} B_i \xrightarrow{\beta_i} C_i \longrightarrow 0$$

is exact. Then we get a diagram with exact rows

$$\begin{array}{ccccccccc} 0 & \longrightarrow & \prod A_i & \longrightarrow & \prod B_i & \longrightarrow & \prod C_i & \longrightarrow & 0 \\ & & \downarrow F & & \downarrow F & & \downarrow F & & \\ 0 & \longrightarrow & \prod A_i & \longrightarrow & \prod B_i & \longrightarrow & \prod C_i & \longrightarrow & 0 . \end{array}$$

The snake lemma now gives an exact sequence

4.23 $$0 \to \varprojlim A_i \to \varprojlim B_i \to \varprojlim C_i \to \varprojlim{}^{(1)} A_i \to \varprojlim{}^{(1)} B_i \to \varprojlim{}^{(1)} C_i \to 0\,.$$

In topology inverse limits often appear via the skeleton filtration of an infinite CW-complex. Suppose F is a spectrum and $F^i(X) = [X, F_i]$ the associated generalized cohomology theory. Then we have from [94]

LEMMA 4.24 (Milnor). *There is a short exact sequence*

$$0 \to \varprojlim_n{}^{(1)} F^{i-1}(X^{(n)}) \to F^i(X) \to \varprojlim_n F^i(X^{(n)}) \to 0\,.$$

The elements of $\lim^{(1)} F^{i-1}(X^{(n)})$ are called *phantom maps* from X to F_i; they are the homotopy classes of maps not detected on any finite subcomplex of X.

Don Anderson has given a convenient criteria for the nonexistence of phantom maps. Recall the spectral sequence ([45]) of the skeleton filtration

$$E_2^{p,q}(X) = H^p(X; F^q(pt))$$

$$E_\infty(X) = E_0(F^*(X))\,.$$

LEMMA 4.25 ([148], [149]). *Suppose in the spectral sequence above that for each* (p, q) *with* $p+q = n$

$$d_r : E_r^{p,q}(X) \to E_r^{p+r, q-r+1}(X)$$

is zero for almost all r. *Then there are no phantom maps from* X *to* F_n.

In particular, if the coefficients $\pi_i(F)$ are finite groups, then there can be no phantom maps.

LEMMA 4.26. *There is a map* $\sigma: G/PL \to BO[½]$ *which on each skeleton* X *of* G/PL *satisfies*

$$\mathrm{eval}(\sigma | X) = \sigma_0 \mid KO_0(X; \mathbf{Z}[½])\,.$$

Proof. There are exact sequences

$$\varprojlim \widetilde{KO}^0(X_i; Z[\tfrac{1}{2}]) \to \varprojlim \mathrm{Hom}(\widetilde{KO}_0(X_i); Z[\tfrac{1}{2}]) \to \varprojlim{}^{(1)} \mathrm{Ext}(\widetilde{KO}_0(SX_i); Z[\tfrac{1}{2}])$$

$$0 \to \varprojlim{}^{(1)} \widetilde{KO}^0(SX_i; Z[\tfrac{1}{2}]) \to [G/PL, BO[\tfrac{1}{2}]] \to \varprojlim \widetilde{KO}^0(X_i; Z[\tfrac{1}{2}]) \to 0$$

where $X_1 \subset X_2 \subset \cdots$ are the skeletons of G/PL. Indeed, the first sequence is a consequence of 4.16 and 4.23 and the second is Milnor's lemma 4.24. Since the Ext-term is finite the right hand side of the first sequence vanishes. This completes the proof.

LEMMA 4.27. *Let* $K = K_4 + K_8 + \cdots$ *be the graded class from* 4.9. *Then* $ph(\sigma) = K$.

Proof. Suppose inductively that

$$ph_{4i}(\sigma) = K_{4i} \quad \text{for} \quad i < n$$

and consider the diagram

$$\begin{array}{ccccccc}
g: S^{4n+4k} & \xrightarrow{\xi} & G/PL^+ \wedge MSO_{4k} & \xrightarrow{\sigma \wedge \Delta} & BO^{\oplus}[\tfrac{1}{2}] \wedge BO^{\oplus}[\tfrac{1}{2}] & \xrightarrow{\otimes} & BO^{\oplus}[\tfrac{1}{2}] \\
\uparrow & & \uparrow & & & & \\
M^{4n} & \xrightarrow{f \times \nu(M)} & G/PL \times BSO_{4k} & & & &
\end{array}$$

where $BO^{\oplus}[\tfrac{1}{2}] = BO[\tfrac{1}{2}] \times (0) \subset BO[\tfrac{1}{2}] \times Z[\tfrac{1}{2}]$ and where ξ and $\{M, f\}$ represent the same element in $\Omega_{4n}(G/PL)$ (cf. 1.25). The composite

$$S^{4n+4k} \longrightarrow (G/PL)^+ \wedge MSO_{4k} \xrightarrow{1 \wedge \Delta} (G/PL)^+ \wedge BO^{\oplus}[\tfrac{1}{2}]$$

represents $\delta(\{M, f\})$. Since

$$ph: \widetilde{KO}^0(S^{4n+4k}; Z[\tfrac{1}{2}]) \to H^{4n+4k}(S^{4n+4k}; Z[\tfrac{1}{2}])$$

maps the chosen generator a^{n+k} onto the cohomology generator, we have

$$\sigma\{M, f\} = \langle g^*(ph_{4n+4k}), [S^{4n+4k}] \rangle$$

where g is the composition above,

$$g = \otimes \circ (\sigma \wedge \Delta) \circ \xi .$$

A computation using the induction hypothesis and the multiplicativity of the Pontrjagin character then leads to

$$\sigma_0\{M, f\} = \langle f^*(K) \cdot \mathcal{L}(M), [M] \rangle + \langle f^*(ph_{4n}(\sigma) - 8K_{4n}), [M] \rangle .$$

Hence by 4.9 $ph_{4n}(\sigma) = K_{4n}$.

THEOREM 4.28 (Sullivan). *The map* σ *defines a homotopy equivalence*

$$\sigma : G/PL[\tfrac{1}{2}] \to BO[\tfrac{1}{2}] .$$

Proof. The spaces in question have the homotopy types of CW complexes, so it suffices to check on homotopy groups. The generator $\iota_{4n} \in \pi_{4n}(G/PL[\tfrac{1}{2}])$ is characterized by

$$s_I\{S^{4n}, \iota_{4n}\} = 1 \qquad (\text{or } 2 \text{ if } n = 1) ,$$

But

$$\begin{aligned} s_I\{S^{4n}, \iota_{4n}\} &= \langle \iota_{4n}^*(K_{4n}), [S^{4n}] \rangle \\ &= \langle \iota_{4n}^*(ph_{4n}(\sigma)), [S^{4n}] \rangle . \end{aligned}$$

Hence $\sigma_*([\iota_{4n}]) = a^n$ where $a^n \in \pi_{4n}(BO[\tfrac{1}{2}])$ is the generator.

E. *The* H-*space structure of* G/PL

A priori the construction of $\sigma: G/PL \to BO[\tfrac{1}{2}]$ in 4.26 only gave a well-defined homotopy class on the finite skeletons. However, any such σ defines a homotopy equivalence, and it is well known that there are no phantom maps from $BO[\frac{1}{2}]$ to itself, so we do in fact have a well-defined homotopy class $\sigma \in [G/PL, BO[\frac{1}{2}]]$. But there is a much stronger result, which together with 4.27 also determines the H-space structure of $G/PL[\frac{1}{2}]$.

THEOREM 4.29. *The Pontrjagin character defines an injection*

$$[BSO, BO] \to H^*(BSO; \mathbf{Q}) .$$

The proof of 4.29 is based on the connection between representation theory and K-theory (see [13], [149]), which we briefly recall.

Let G be a compact (connected) Lie group and P a principal G-bundle over the finite CW-complex X. To each virtual orthogonal representation $V-W$ of G we assign the virtual vector bundle $P \times_G V - P \times_G W$; this defines a ring homomorphism

$$A_X : RO(G) \to KO(X) .$$

Let $IO(G)$ denote the augmentation ideal in $RO(G)$ (of virtual representations of degree 0). Elements in the image of $IO(G)$ vanish on the zero skeleton of X, so $A_X(IO(G)^n) = 0$ if $n > \dim X$. Hence, if $RO(G)^\wedge$ denotes the completion of $RO(G)$,

$$RO(G)^\wedge = \varprojlim_k RO(G)/IO(G)^k ,$$

we have a homomorphism

$$A : RO(G)^\wedge \to \varprojlim_n [n\text{-skeleton of } BG, BO] .$$

From [13] we have

THEOREM 4.30. (i) *There are no phantom maps from* BG *to* BO.

(ii) $A : IO(G)^\wedge \to [BG, BO]$ *is an isomorphism.*

We get 4.29 from 4.30 by specializing to $G = SO(2n+1)$. Indeed, recall that

$$RO(SO(2n+1)) = \mathbf{Z}[\gamma_1, \gamma_2, \cdots, \gamma_n] ,$$

where $\gamma_i = \gamma^i(V - \dim V)$*) and V is the standard representation of $SO(2n+1)$ on R^{2n+1} (see e.g. [5]). The augmentation ideal $IO(SO(2n+1))$ is the usual augmentation ideal in the stated polynomial algebra, so

$$RO(SO(2n+1))^\wedge = Z[[\gamma_1, \gamma_2, \cdots, \gamma_n]],$$

the corresponding power series ring. From 4.30 we then have

$$[BSO(2n+1), BO] = Z[[\gamma_1, \gamma_2, \cdots, \gamma_n]].$$

Finally, each finite skeleton of BSO is contained in some $BSO(2n+1)$, so

$$[BSO, BO] = Z[[\gamma_1, \gamma_2, \cdots]].$$

Now, the proof of 4.29 is direct: $[BSO, BO]$ maps injectively into $[BSO, BO] \otimes Q \cong [BSO; BO[Q]]$ where $BO[Q]$ denotes the rational type of BO. But

$$ph : BO[Q] \to \prod_{n \geq 1} K(Q, 4n)$$

is a homotopy equivalence.

Let $BO^{\otimes} = BO \times (1) \subset BO \times Z$, organized into an H-space by tensor product of virtual bundles of dimension 1.

COROLLARY 4.31. $\hat{\sigma} = 1 + 8\sigma : G/PL[\frac{1}{2}] \to BO^{\otimes}[\frac{1}{2}]$ *is an H-equivalence.*

(Indeed, $ph(\hat{\sigma}) = 1 + 8K = \hat{j}^*(\mathcal{L}_{PL})$ where $\hat{j} : G/PL \to BSPL$. But $\hat{j}^*(\mathcal{L}_{PL})$ is multiplicative so $\hat{\sigma}$ must be an H-map.)

The H-structure of $G/PL[2]$ is harder. In 4.9 we constructed primitive fundamental classes $K_{4n-2} \in H^{4n-2}(G/PL; Z/2)$ and classes $K_{4n} \in F^{4n}(G/PL; Z_{(2)})$ such that

$$\psi(K_{4n}) = 1 \otimes K_{4n} + K_{4n} \otimes 1 + 8 \sum K_{4i} \otimes K_{4(n-i)}$$

---

*) Here γ^i denotes the i'th γ-power, see Chapter 9.B or [12] for a definition.

in $F^*(G/PL \times G/PL; Z_{(2)})$. But there is no easy way of lifting the K_{4n} to 'good' classes in $H^*(G/PL; Z_{(2)})$. We content ourselves to quote from [91] and [103].

THEOREM 4.32.*) *There are classes* $\tilde{K}_{4n} \in H^{4n}(G/PL; Z_{(2)})$ *reducing to the classes* K_{4n} *from* 4.9 *and with diagonal*

$$\psi(\tilde{K}_{4n}) = 1 \otimes \tilde{K}_{4n} + \tilde{K}_{4n} \otimes 1 + 8 \sum \tilde{K}_{4i} \otimes \tilde{K}_{4(n-i)} .$$

For many calculations the generators $\tilde{K}_{4n}$ are unsuitable. However, due to the factor 8 in $\bar{\psi}(\tilde{K}_{4n})$ it is possible to choose new primitive**) generators.

4.33
$$k_{4n} = \tilde{K}_{4n} + 4D_{4n}, \quad n \geq 1$$

where D_{4n} is a polynomial in the $\tilde{K}_4, \cdots, \tilde{K}_{4n-4}$ with $Z_{(2)}$ coefficients (see [78]). In conclusion, we have

---

*)The classes K_{4n} are specified by formulae for surgery obstructions of surgery problems over '$Z/2^r$-manifolds', analogous to the formulas in 4.9 but quite a bit more complicated.

The classes defined in [91] and [103] are not identical. Indeed, the difference between the two 'total' classes was calculated in [34]. It is the graded class

$$\beta\Big(\sum_{i \geq 1} Sq^{2^i}\Big) Sq^1\Big(\sum_{n \geq 1} K_{4n-2}\Big) .$$

where β denotes the Bockstein operator, $\beta: H^*(\ ; Z/2) \to H^*(\ ; Z_{(2)})$.

The mod. 2 reduction of $\tilde{K}_4 \in H^4(G/PL; Z_{(2)})$ is K_2^2. In $H^4(G/PL; Z/2)$ a new class K_4' appears and

$$\psi(K_4') = 1 \otimes K_4' + K_4' \otimes 1 + K_2 \otimes K_2 .$$

We refer the reader to [35] for further details.

**)A class $k \in H^*(G/PL; Z_{(2)})$ is called primitive if $\mu^*(k) = 1 \otimes k + k \otimes 1$ where $\otimes$ denotes the exterior product $\otimes: H^*(G/PL; Z_{(2)}) \otimes H^*(G/PL; Z_{(2)}) \to H^*(G/PL \times G/PL; Z_{(2)})$.

THEOREM 4.34 (Sullivan). *There are H-equivalences*

(a) $G/PL[2] \simeq \Omega E_3 \times \prod_{n>1} K(Z_{(2)}, 4n) \times K(Z/2, 4n-2)$

(b) $G/PL[\frac{1}{2}] \simeq BO^{\otimes}[\frac{1}{2}]$

where E_3 *is the fiber in*

$$E_3 \longrightarrow K(Z/2, 3) \xrightarrow{\beta Sq^2} K(Z_{(2)}, 6) .$$

We can assemble the local data in terms of a Cartesian square in the category of H-spaces to obtain the actual integral type of G/PL

4.35
$$\begin{array}{ccc} G/PL & \longrightarrow & G/PL[2] \\ \downarrow{\scriptstyle 1+8\sigma} & & \downarrow{\scriptstyle 8K} \\ BO^{\otimes}[\frac{1}{2}] & \xrightarrow{ph} & \Pi K(Q, 4n) . \end{array}$$

Alternatively, we can describe G/PL as the fiber in the sequence

$$G/PL \longrightarrow G/PL[2] \times G/PL[\tfrac{1}{2}] \xrightarrow{8K \times ph((1+8\sigma)^{-1})} \Pi K(Q, 4n) .$$

REMARK 4.36. The results for G/TOP are quite similar. It follows from 2.25 that

4.37
$$G/TOP[2] \simeq \prod_{n>0} K(Z_{(2)}, 4n) \times K(Z/2, 4n-2) .$$

Indeed, the classes $\tilde{K}_{4n} \in H^{4n}(G/PL; Z_{(2)})$ and $K_{4n-2} \in H^{4n-2}(G/PL; Z/2)$ come from cohomology classes on G/TOP also denoted $\tilde{K}_{4n}, K_{4n-2}$. Their diagonals are again given by

$$\psi(\tilde{K}_{4n}) = 1 \otimes \tilde{K}_{4n} + \tilde{K}_{4n} \otimes 1 + 8 \sum \tilde{K}_{4i} \otimes \tilde{K}_{4(n-i)}$$

$$\psi(K_{4n-2}) = 1 \otimes K_{4n-2} + K_{4n-2} \otimes 1$$

and the equivalence in 4.37 is given by $\sum \tilde{K}_{4n} \times \sum K_{4n-2}$ (or $\sum k_{4n} \times \sum K_{4n-2}$, cf. 4.33).

The classes $\tilde{K}_{4n}, K_{4n-2}$ on G/TOP are related to the surgery obstructions for topological degree 1 normal maps by the formulae in 4.9. Precisely, if $n \neq 4$ and $f: \hat{M}^n \to M^n, \hat{f}: \nu(\hat{M}) \to \xi$ is the surgery problem associated to $\gamma: M^n \to G/TOP$ then its index and Kervaire obstructions are given in 4.9 by substituting $\tilde{K}_{4n}$ for K_{4n}.*)

For $n = 4$ topological transversality is not known. However, there still is a mapping $s_I: [M^4, G/TOP] \to \mathbf{Z}$: if $\gamma: M^4 \to G/TOP$, consider $\gamma \circ p_2: CP^2 \times M^4 \to M^4 \to G/TOP$ and define $s_I(M^4, \gamma) = s_I(CP^2 \times M^4, \gamma \circ p_2)$. Then $s_I(M^4, \gamma) = \langle \gamma^*(\tilde{K}_4), [M^4] \rangle$.

Finally, the diagram 4.35 remains a Cartesian square when we replace G/PL by G/TOP.

---

*) If M^n is not differentiable but merely a PL or topological manifold then the formulae of 4.9 remain valid if we substitute for $\mathcal{L}(\tau(M))$ the class $\mathcal{L}_{PL}(\tau(M))$ or $\mathcal{L}_{TOP}(\tau(M))$. $\mathcal{L}_{PL}$ was defined in 4.18 and $\mathcal{L}_{TOP} \in F^*(BTOP; \mathbf{Z}_{(2)})$ can be defined quite analogously (cf. [103]).

CHAPTER 5

THE HOMOTOPY STRUCTURE OF MSPL[½] AND MSTOP[½]

In the previous chapter we saw how the KO(; Z[½]) orientability of oriented vector bundles allowed us to reinterpret geometric invariants of the smooth bordism groups of certain spaces in terms of characteristic classes. In this section we go on to review Sullivan's analysis of the classifying spaces BSTOP[½] and BSPL[½] and their associated Thom spectra. The starting point is the construction of a KO(; Z[½]) orientation of stable PL and TOP bundles. This leads to a determination of the odd-local homotopy types of BSPL and BSTOP modulo a largely unknown space, BcokJ, whose homotopy is the cokernel of the J-homomorphism in the stable homotopy groups of spheres. The cohomology groups $H^*(\mathrm{BcokJ})$ are known ([74], [86]) but very complicated (and not needed for our purpose) whereas the K-theory of BcokJ is trivial ([56]): $\widetilde{KO}^*(\mathrm{BcokJ}; \mathbf{Z}[½]) = 0$. Finally, combining these results with the (now verified) Adams' conjecture one obtains a similar analysis of the Thom spectra MSPL[½] and MSTOP[½].

A. *The* KO-*orientation of* PL-*bundles away from* 2

Let $\Omega_*(\mathrm{MSPL})$ be the smooth bordism of the spectrum MSPL away from 2,

$$\Omega_n(\mathrm{MSPL}) = \varinjlim_r \Omega^{SO}_{n+r}(\mathrm{MSPL}_r) \otimes \mathbf{Z}[½]$$

and let

5.1 $$I\colon \Omega_*(\mathrm{MSPL}) \to \mathbf{Z}[½]$$

be the homomorphism which to a singular manifold

$$f\colon Q^{4n+r} \to \mathrm{MSPL}_r\,,$$

with f transverse to the zero section, associates the index of the codimension r submanifold, $M^{4n} = f^{-1}(\mathrm{BSPL}_r)$. As in Chapter 4.D., I factors over the surjection

$$\delta : \Omega_{4*}(\mathrm{MSPL}) \to \mathrm{KO}_0(\mathrm{MSPL}; \mathbf{Z}[\tfrac{1}{2}]) .$$

The universal coefficient theorem 4.16 can be extended to spectra ([147]) and gives an exact sequence

$$0 \to \mathrm{Ext}(\widetilde{\mathrm{KO}}_{-1}(\mathrm{MSPL}),\mathbf{Z}[\tfrac{1}{2}]) \to \widetilde{\mathrm{KO}}^0(\mathrm{MSPL};\mathbf{Z}[\tfrac{1}{2}]) \to \mathrm{Hom}(\widetilde{\mathrm{KO}}_0(\mathrm{MSPL}),\mathbf{Z}[\tfrac{1}{2}]) \to 0 .$$

In particular there is a class

5.2 $$\Delta_{\mathrm{PL}} \in \widetilde{\mathrm{KO}}(\mathrm{MSPL}; \mathbf{Z}[\tfrac{1}{2}])$$

which defines the homomorphism I in 5.1. (We will see in 5.D below that the term $\mathrm{Ext}(\widetilde{\mathrm{KO}}_{-1}(\mathrm{MSPL}); \mathbf{Z}[\tfrac{1}{2}])$ vanishes, and Δ_{PL} in 5.2 is in fact unambigously defined.)

LEMMA 5.3. *Under the natural map*

$$\mathrm{Mi} : \mathrm{MSO} \to \mathrm{MSPL}$$

the class Δ_{PL} *is mapped onto the orientation class* $\Delta = \Delta_{\mathrm{SO}}$ *of* 4.14.

Proof. Since $\widetilde{\mathrm{KO}}_{-1}(\mathrm{MSO}; \mathbf{Z}[\tfrac{1}{2}]) = 0$ it suffices to check that $(\mathrm{Mi})^*(\Delta_{\mathrm{PL}})$ and Δ_{SO} define the same element in $\mathrm{Hom}(\widetilde{\mathrm{KO}}_0(\mathrm{MSO}), \mathbf{Z}[\tfrac{1}{2}])$. Since δ is onto, it is enough to see that $(\mathrm{Mi})^*(\Delta_{\mathrm{PL}})$ and Δ_{SO} determine the same homomorphism from $\Omega_*(\mathrm{MSO})$ to $\mathbf{Z}[\tfrac{1}{2}]$.

An element $\{Q^{4n+4r}, f\} \in \Omega_{4n+4r}(\mathrm{MSO}_{4r})$ is represented by a map

$$a : S^{4n+4r+4s} \to \mathrm{MSO}_{4s} \wedge \mathrm{MSO}^+_{4r}$$

$(r >> n, s >> n)$ and by transversality we get a diagram

$$\begin{array}{ccc}
S^{4n+4r+4s} & \xrightarrow{\ a\ } & MSO_{4s} \wedge MSO^{+}_{4r} \\
\cup & & \cup \\
Q^{4n+4r} & \xrightarrow{(\nu_Q, f)} & BSO_{4s} \times MSO_{4r} \\
\cup & & \cup \\
M^{4n} & \xrightarrow{\nu_Q|M \oplus \nu(M:Q)} & BSO_{4s} \times BSO_{4r}
\end{array}$$

Now,

$$<Mi^*(\Delta_{PL}), [Q^{4n+4r}, f]> \; = \; \mathrm{Index}(M^{4n})$$

and from 4.19,

$$\begin{aligned}
\mathrm{Index}(M) &= <\nu_M^*(\mathcal{L}^{-1}), [M]> \\
&= <\nu(M:Q)^*(\mathcal{L}^{-1}) \cdot (\nu_Q|M)^*(\mathcal{L}^{-1}), [M]> \\
&= <f^*(\mathcal{L}^{-1} \cdot U) \cdot \nu_Q^*(\mathcal{L}^{-1}), [Q]> \\
&= <a^*(\mathcal{L}^{-1} U \otimes \mathcal{L}^{-1} U), [S^{4n+4r+4s}]> .
\end{aligned}$$

On the other hand, the definition of the Kronecker pairing between K-homology and K-cohomology (see Chapter 4.B) implies that $\Delta_{SO}(\{Q,f\})$ is the homotopy class of the composite

$$g: S^{4n+4r+4s} \longrightarrow MSO_{4s} \wedge MSO^{+}_{4r} \xrightarrow{\Delta \wedge \Delta} BSO \wedge BSO \xrightarrow{\otimes} BSO[1/2] .$$

Since the Pontrjagin character evaluates to 1 on the chosen generator of $\pi_{4i}(BSO[1/2])$ and because $ph\Delta_{SO} = \mathcal{L}^{-1} \cdot U$ we have

$$\begin{aligned}
<\Delta_{SO}, \{Q^{4n+4r}, f\}> &= <ph(g), [S^{4n+4r+4s}]> \\
&= <a^*(\mathcal{L}^{-1} U \otimes \mathcal{L}^{-1} U), [S^{4n+4r+4s}]>
\end{aligned}$$

and the lemma follows.

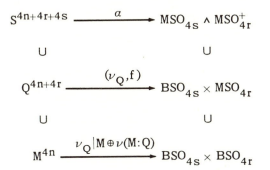

COROLLARY 5.4. $ph\Delta_{PL} = (\mathcal{L}_{PL})^{-1} \cdot U_{PL}$ *where* U_{PL} *is the cohomology Thom class. In particular* Δ_{PL} *is an orientation class for* MSPL.

Proof. We saw in Chapter 4 that $\mathcal{L}_{PL}$ restricts to $\mathcal{L}$ under the natural map $i: BSO \to BSPL$. The fiber is the space PL/O whose homotopy groups are finite. Hence i is a rational equivalence and the result follows from 4.14.

B. *The splitting of* p*-local* PL*-bundles,* p *odd*

Let ξ be a 4r-dimensional PL-bundle over X ($4r >> \dim X$) and let $\Delta_{PL}(\xi) \in \widetilde{KO}(M(\xi); Z[½])$ be the orientation class constructed above. We mimic a standard procedure for vector bundles and define a characteristic class for each integer k

$$\rho_L^k(\xi) \in 1 + \widetilde{KO}(X; Z[½, 1/k]) .$$

Indeed, let $\rho_L^k(\xi)$ be the unique class so that

$$(*) \qquad \rho_L^k(\xi) \cdot \Delta_{PL}(\xi) = \frac{1}{k^{2r}} \psi^k(\Delta_{PL}(\xi))$$

where ψ^k denotes the Adams operation [9]. Universally we get a mapping

$$5.5 \qquad \rho_L^k : BSPL \to BO^{\otimes}[½, 1/k] .$$

The Pontrjagin character of ρ_L^k can be calculated from (*). From 5.12 and 5.22 below we have that $[BSPL, BO[½]] \to [BSPL, BO] \otimes Q$ is injective. Hence the homotopy class of ρ_L^k in 5.5 is uniquely determined.

In the rest of this chapter p will be an odd prime unless otherwise indicated, and k will be a positive number which reduces to a generator of the group of units in Z/p^2 : $k^{p-1} \not\equiv 1 \pmod{p^2}$.

DEFINITION 5.6. The homotopy theoretical fiber of $\rho_L^k: BSPL[p] \to BSO^{\otimes}[p]$ is denoted $BcokJ_p$.

LEMMA 5.7. *Let* $\hat{\sigma} = 1 + 8\sigma$ *where* σ *was characterized in* 4.26. *Then*

$$\begin{array}{ccc} G/PL & \xrightarrow{\hat{\sigma}} & BSO^{\otimes}[1/2, 1/k] \\ \downarrow{\scriptstyle \hat{j}} & & \downarrow{\scriptstyle 1/\psi^k} \\ BSPL & \xrightarrow{\rho_L^k} & BSO^{\otimes}[1/2, 1/k] \end{array}$$

is homotopy commutative.

Proof. As $\hat{\sigma}$ is a homotopy equivalence away from 2 and k, and since homotopy classes of endomorphisms of BSO are distinguished by their Pontrjagin character by 4.29 it suffices to check that

$$ph(\hat{\sigma}/\psi^k\hat{\sigma}) = ph(\rho_L^k \circ \hat{j}) .$$

We evaluate the Pontrjagin character of ρ_L^k from the defining equation on 4r dimensional bundles

$$\rho_L^k \cdot \Delta_{PL} = \frac{1}{k^{2r}} \psi^k(\Delta_{PL}) .$$

But $ph(\psi^k(x)) = \psi_H^k(ph(x))$ where

$$\psi_H^k : H^{4i}(\ ; Q) \to H^{4i}(\ ; Q)$$

is multiplication by k^{2i}. Since $ph\,\Delta_{PL} = \mathcal{L}^{-1} \cdot U_{PL}$ where U_{PL} has dimension 4r we have

$$ph\,\rho_L^k = \psi_H^k(\mathcal{L}^{-1})/\mathcal{L}^{-1} = \mathcal{L}/\psi_H^k(\mathcal{L})$$

and the result follows from 4.22 and 4.27.

In [2] Adams considered a map

$$\rho_A^k : BSO^{\oplus}[p] \to BSO^{\otimes}[p]$$

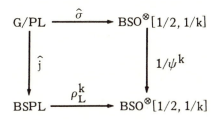

and proved that it induces isomorphisms on homotopy groups in dimensions $2(p-1)n$. The Pontrjagin character of ρ_A^k is

$$\mathrm{ph}\,\rho_A^k = \mathfrak{A}$$

where $\mathfrak{A}$ is the genus with characteristic power series

$$S(\mathfrak{A}) = \frac{e^{kx/2} - e^{-kx/2}}{k(e^{x/2} - e^{-x/2})} .$$

A check of Pontrjagin characters shows

$$\rho_L^k(x) = \rho_A^k(\psi^2(2x) - \psi^4(x))$$

and since the mapping $2\psi^2 - \psi^4 : BSO \to BSO$ induces multiplication by $2^{(p-1)n}(2-2^{(p-1)n})$ on homotopy in dimension $2(p-1)n$ we get

LEMMA 5.8. *The maps*

$$\rho_L^k : BSO[p] \to BSO[p]$$

$$\psi^k - 1 : BSO[p] \to BSO[p]$$

induce isomorphisms on the homotopy groups in dimension $2(p-1)n$ *and* 2ℓ *for* $\ell \not\equiv 0(p-1)$, *respectively.*

D. Anderson constructed a splitting of the p-local space $BSO^{\oplus}[p]$ (see e.g. [3])

5.9 $$BSO^{\oplus}[p] \simeq BSO_{(1)} \times \cdots \times BSO_{(m)}, \quad m = \frac{p-1}{2} .$$

The homotopy groups of $BSO_{(i)}$ are concentrated in dimensions congruent to $4i-4 \pmod{(2p-2)}$. Each of the factors is an H-space and the splitting is in the category of H-spaces. Also $BSO^{\oplus}[p]$ and $BSO^{\otimes}[p]$ are equivalent as H-spaces. Indeed Atiyah and Segal in [14] exhibited an H-equivalence

5.10 $$\delta : BSO^{\oplus}[p] \xrightarrow{\simeq} BSO^{\otimes}[p] .$$

(We review the construction of δ in Chapter 9.B.) Let $\pi_1, \pi_1^{\perp}$: BSO[p] → BSO[p] be the composites

$$\pi_1 : BSO[p] \xrightarrow{\text{proj}} BSO_{(1)} \xrightarrow{\text{incl}} BSO[p]$$

$$\pi_1^{\perp} : BSO[p] \xrightarrow{\text{proj}} BSO_{(1)}^{\perp} \xrightarrow{\text{incl}} BSO[p]$$

where $BSO_{(1)}^{\perp} = BSO_{(2)} \times \cdots \times BSO_{(m)}$. Let γ_p be the H-map

5.11 $$\gamma_p : BSO^{\oplus}[p] \xrightarrow{\text{diag}} BSO^{\oplus}[p] \times BSO^{\oplus}[p] \xrightarrow{\pi_1 \times \pi_1^{\perp}} BSO[p] \times BSO[p]$$

$$\xrightarrow{1 \times \hat{\sigma}^{-1}\delta} BSO^{\oplus}[p] \times G/PL[p] \xrightarrow{\tilde{j} \times \hat{j}} BSPL[p] \times BSPL[p] \xrightarrow{\oplus} BSPL[p] .$$

THEOREM 5.12 (Sullivan). *There is a splitting*

$$BSPL[p] \simeq BSO^{\oplus}[p] \times BcokJ_p$$

at each odd prime p.

Proof. It suffices to see that γ_p splits ρ_L^k, that is,

$$\rho_L^k \circ \gamma_p : BSO[p] \to BSO^{\otimes}[p]$$

is a homotopy equivalence; since we can then define an equivalence from $BSO^{\oplus}[p] \times BcokJ_p$ into BSPL[p] as the sum of γ_p and the inclusion of $BcokJ_p$. We check on homotopy groups: In degrees $2(p-1)n$, $(\rho_L^k \circ \gamma_p)_*$ is an isomorphism by 5.8 and in dimensions $2m$ with $m \not\equiv 0(p-1)$, $\rho_L^k \circ \gamma_p$ induces essentially the same map as $(\psi^k - 1)$, that is, multiplication by $k^m - 1$. But $(p, k^m - 1) = 1$ in our situation. This completes the proof.

C. *The homotopy types of* G/O[p] *and* SG[p]

In order to further clarify the homotopy structure of BSPL and MSPL we digress to give a discussion of G/O. The Adams conjecture proved in [18], [111], [136] implies a diagram

5.13
$$\begin{array}{ccc} BSO[p] & \xrightarrow{\alpha_p} & G/O[p] \\ & \underset{\psi^k - 1}{\searrow} & \downarrow \\ & & BSO[p] \to BSG[p] \end{array}$$

for each prime p. Here we take $k \equiv \pm 3 \pmod 8$ (and usually $k = 3$) if $p = 2$ and otherwise (as usual) k is a generator of the group of units in $\mathbf{Z}/p^2$.

The map α_p turns out to be a split injection. We first construct for each prime p an H-map

$$\beta_p : G/O[p] \to BSO^{\otimes}[p] .$$

Then in 5.18 below we prove that $\beta_p \circ \alpha_p$ is a homotopy equivalence of H-spaces. The construction of β_p is somewhat different when p is an odd prime and when $p = 2$. We begin with the case p odd.

Previously we have used the H-space splitting of $BSO^{\oplus}[p]$ into a product, $BSO^{\oplus}[p] \simeq BSO_{(1)} \times BSO_{(1)}^{\perp}$. We use the Atiyah-Segal H-equivalence $\delta_p : BSO^{\oplus}[p] \to BSO^{\otimes}[p]$ to get an analogous multiplicative splitting

$$BSO^{\otimes}[p] \simeq BSO^{\otimes}_{(1)} \times (BSO^{\otimes}_{(1)})^{\perp} .$$

Let $\overline{r}, \overline{s}$ be the compositions

$$\overline{r}: G/O[p] \xrightarrow{r} BSO^{\oplus}[p] \xrightarrow{\delta_p} BSO^{\otimes}[p] \xrightarrow{proj} (BSO^{\otimes}_{(1)})^{\perp}$$

$$\overline{s}: G/O[p] \xrightarrow{s} G/PL[p] \xrightarrow{\hat{\sigma}} BSO^{\otimes}[p] \xrightarrow{proj} BSO^{\otimes}_{(1)}$$

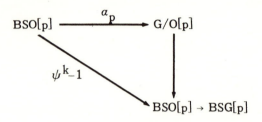

where r and s are the natural inclusions, and define β_p as the composition

5.14 $$\beta_p : G/O[p] \xrightarrow{\text{diag.}} G/O[p] \times G/O[p] \xrightarrow{\bar{s}\times\bar{r}} BSO^{\otimes}[p] .$$

The construction of β_2 is analogous, but in order to make it precise we first give a new definition of $\hat{\sigma}: G/PL[p] \to BSO^{\oplus}[p]$. (Cf. 4.31.) Since G/PL is the fiber of the natural map

$$\pi : BSPL \to BSG$$

it is the classifying space for pairs (E, t) consisting of a stable PL-bundle E (over X) and a proper fiber homotopy equivalence

$$t: E \to \varepsilon_X$$

where ε_X is a stably trivial bundle with a given trivialization ($\varepsilon_X = X \times R^N$). The bundles E and ε_X have specific KO-orientations at odd primes by 5.2, and we can define a characteristic class $e_L(E,t) \in 1 + \widetilde{KO}(X) \otimes Z_{(p)}$ by

$$\Delta_{PL}(E) \cdot e_L(E,t) = t^*(\Delta_{PL}(\varepsilon_X)) .$$

Since G/PL is the universal object for pairs (E, t) as above the characteristic class $e_L(E,t)$ determines a map

$$e_L : G/PL \to BSO[p] .$$

LEMMA 5.15. *At every odd prime* p, $e_L = \hat{\sigma}$.

Proof. It suffices to check that $ph(e_L) = ph(\hat{\sigma})$ (cf. 4.29), and from the defining equation we have

$$\hat{j}^* ph(\Delta_{PL}) \cdot ph(e_L) = \hat{j}^*(U_{PL}) ,$$

where $\hat{j}: G/PL \to BSPL$ is the natural map.

Thus 4.22 and 5.4 give

$$\mathrm{ph}(e_L) = 1 + 8K .$$

But $\mathrm{ph}(\sigma) = K$ by 4.27 and $\hat{\sigma} = 1 + 8\sigma$ so the lemma follows.

We now construct $\beta_2 : G/O[2] \to BSO^{\otimes}[2]$. The space G/O classifies pairs (E, t) as above but with the difference that E is now a stable vector bundle. Oriented vector bundles are not in general KO-oriented (at the prime 2). However, the natural map $r: G/O \to BSO$ lifts to BSpin (since the Stiefel-Whitney classes in $H^*(BSO; \mathbf{Z}/2)$ come from $H^*(BSG; \mathbf{Z}/2)$, and BSpin is the homotopy fiber of $BSO \xrightarrow{w_2} K(\mathbf{Z}/2, 2)$). Moreover, as G/O is simply connected the homotopy class of a lifting $\bar{r}: G/O \to BSpin$ is unique. Hence for a pair (E, t), E admits a unique Spin structure and therefore a well-defined Thom class [10],

$$\Delta_A(E) \in \widetilde{KO}(ME) .$$

We proceed as above and define a characteristic class $e_A(E, t)$ in $1 + \widetilde{KO}(X)$,

$$\Delta_A(E) \cdot e_A(E, t) = t^*(\Delta_A(\varepsilon_X)) .$$

The natural Thom class Δ_A is exponential, $\Delta_A(E \oplus F) = \Delta_A(E) \cdot \Delta_A(F)$, so we get an H-map

$$e_A : G/O \to BSO^{\otimes}$$

and we let β_2 be the localization at 2 of e_A.

From ([2], J(X) II) we have that $\mathrm{ph}\, \Delta_A = \mathfrak{A} \cdot U_H$, where $\mathfrak{A}$ is the genus with characteristic power series

$$S(\mathfrak{A}) = \frac{e^{x/2} - e^{-x/2}}{x} ,$$

so $\mathrm{ph}(e_A) = r^*(\mathfrak{A})$.

DEFINITION 5.16. For each prime p, let $\mathrm{cok}\, J_p$ be the homotopy fiber of

$$\beta_p : G/O[p] \to BSO^{\otimes}[p] .$$

The next result justifies the notation in 5.6 and 5.16.

THEOREM 5.17. *For an odd prime* p, $\Omega B\mathrm{cok}J_p \simeq \mathrm{cok}J_p$.

Proof. The diagram in 5.7 induces a map $(\hat{\sigma}_1, \hat{\sigma}, \rho_L^k)$ of homotopy fiberings

$$\begin{array}{ccccc} SG[p] & \longrightarrow & G/PL[p] & \longrightarrow & BSPL[p] \\ \downarrow{\scriptstyle \hat{\sigma}_1} & & \downarrow{\scriptstyle \hat{\sigma}} & & \downarrow{\scriptstyle \rho_L^k} \\ J_p^{\otimes} & \longrightarrow & BSO^{\otimes}[p] & \xrightarrow{1/\psi^k} & BSO^{\otimes}[p] , \end{array}$$

where $J_p^{\otimes}$ is the homotopy fiber of $1/\psi^k$. Since $\hat{\sigma}$ is a homotopy equivalence the fiber of $\hat{\sigma}_1$ is homotopy equivalent to $\Omega B\mathrm{cok}J_p$. On the other hand, we have the map $(\hat{\sigma}_2, \beta_p, \bar{\rho})$ of homotopy fiberings,

$$\begin{array}{ccccc} SG[p] & \longrightarrow & G/O[p] & \longrightarrow & BSO^{\oplus}[p] \\ \downarrow{\scriptstyle \hat{\sigma}_2} & & \downarrow{\scriptstyle \beta_p} & & \downarrow{\scriptstyle \bar{\rho}} \\ J_p^{\otimes} & \longrightarrow & BSO^{\otimes}[p] & \xrightarrow{1/\psi^k} & BSO^{\otimes}[p] . \end{array}$$

Here $\bar{\rho} = \otimes \circ (\rho_L^k \times (1/\psi^k) \circ \delta_p) \circ (\pi_1 \times \pi_1^{\perp}) \circ \mathrm{diag}$, where π_1 and $\pi_1^{\perp}$ are the 'projections' used in 5.11 and $\otimes : BSO^{\otimes}[p] \times BSO^{\otimes}[p] \to BSO^{\otimes}[p]$ is the multiplication in the indicated H-structure.

Since $\bar{\rho}$ is a homotopy equivalence the homotopy theoretic fibers of $\hat{\sigma}_2$ and β_p agree up to homotopy so it suffices to show that $\hat{\sigma}_1 \simeq \hat{\sigma}_2$.

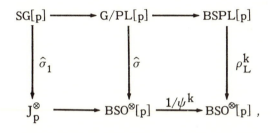

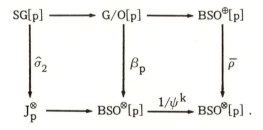

The diagram

$$\begin{array}{ccc} SG[p] & \longrightarrow & G/O[p] \\ \Big\downarrow \hat{\sigma}_1 & & \Big\downarrow \beta_p \\ J_p^{\otimes} & \longrightarrow & BSO^{\otimes}[p] \end{array}$$

homotopy commutes by the definition of β_p, and the quotient $\hat{\sigma}_1/\hat{\sigma}_2$ then lifts to a homotopy class of maps from SG to SO[p]. We claim that $[SG, SO[p]] = 0$.

Indeed, from 3.9 we have that SG is homology equivalent to $B\Sigma_\infty = \varprojlim B\Sigma_n$, and the usual Atiyah-Hirzebruch spectral sequence shows that SG and $B\Sigma_\infty$ have the same KO-groups. (In Chapter 5.F we shall examine this equivalence in greater detail.) In particular

$$[SG, SO[p]] \cong KO^{-1}(B\Sigma_\infty) \otimes \mathbf{Z}_{(p)}\,.$$

Now, $KO^{-1}(B\Sigma_\infty) \otimes \mathbf{Z}_{(p)} \cong \varprojlim KO^{-1}(B\Sigma_n) \otimes \mathbf{Z}_{(p)}$ and since $KO^{-1}(X) \otimes \mathbf{Z}_{(p)}$ is a direct factor of the p-local complex K-group $K^{-1}(X) \otimes \mathbf{Z}_{(p)}$, the assertion follows from Atiyah's result from [149]: $K^{-1}(B\pi) = 0$ for π a finite group.

THEOREM 5.18 (Sullivan). *At each prime* p,

$$G/O[p] \simeq BSO[p] \times \mathrm{cok}J_p$$

$$SG[p] \simeq J_p \times \mathrm{cok}J_p\,,$$

where J_p *is the homotopy theoretical fiber of* $\psi^k - 1\colon BSO^{\oplus}[p] \to BSO^{\oplus}[p]$.

(Each of the spaces above has a natural H-structure, and one might ask if the homotopy equivalences above are H-space equivalences. This is indeed the case for p odd (see e.g. [85]) but neither G/O[2] nor SG[2] splits as H-spaces in the indicated manner ([73]).)

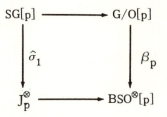

Proof. We give the proof for p an odd prime and leave the somewhat easier case $p = 2$ to the reader. Let $\alpha_p : BSO[p] \to G/O[p]$ be the map given by the Adams conjecture as in 5.13.

We first show that the composition

$$BSO[p] \xrightarrow{\alpha_p} G/O[p] \xrightarrow{s} G/PL[p] \xrightarrow{\hat{\sigma}} BSO^{\otimes}[p]$$

is $1/\rho_L^k$, the inverse of the cannibalistic class ρ_L^k.

By 4.29 it suffices to evaluate Pontrjagin characters, and using 4.22 we see that $ph(\hat{\sigma} \circ s) = r^*(\mathcal{L})$ where $\mathcal{L}$ is the Hirzebruch genus in $H^*(BSO; Q)$, and $r : G/O \to BSO$ is the usual inclusion. By the definition of α_p, $\psi^k - 1 = r \circ \alpha_p$ so we have

$$ph(\hat{\sigma} \circ s \circ \alpha_p) = (\psi^k - 1)^*(\mathcal{L}) .$$

The class $\mathcal{L}$ is exponential so

$$\log(\mathcal{L}) = (\mathcal{L}-1) - \frac{1}{2}(\mathcal{L}-1)^2 + \frac{1}{3}(\mathcal{L}-1)^3 - \cdots$$

is additive, that is, a graded primitive class. On primitive classes $(\psi^k - 1)^*$ is easily calculated,

$$\begin{aligned} (\psi^k-1)^*(\log \mathcal{L}) &= (\psi^k)^*(\log \mathcal{L}) - \log \mathcal{L} \\ &= \log(\psi^k)^*(\mathcal{L}) - \log \mathcal{L} \\ &= \log((\psi^k)^*(\mathcal{L})/\mathcal{L}) . \end{aligned}$$

But induced maps clearly commute with log,

$$(\psi^k-1)^*(\log \mathcal{L}) = \log((\psi^k-1)^*(\mathcal{L}))$$

so we have

$$(\psi^k-1)^*(\mathcal{L}) = (\psi^k)^*(\mathcal{L})/\mathcal{L}$$

which is equal to $ph(1/\rho_L^k)$ (cf. the proof of 5.7). Hence $\sigma \circ s \circ \alpha_p = 1/\rho_L^k$ as claimed.

From 5.8 it now follows that

$$\beta_p \circ \alpha_p : BSO[p] \to BSO[p]$$

induces an isomorphism on homotopy groups, and the composition

$$BSO[p] \times cokJ_p \xrightarrow{\alpha_p \times \text{incl.}} G/O[p] \times G/O[p] \longrightarrow G/O[p]$$

is the required homotopy equivalence.

The splitting of $SG[p]$ is derived from the splitting of $G/O[p]$. Consider the diagram

$$\begin{array}{ccccc}
J_p & \longrightarrow & BSO^{\oplus}[p] & \xrightarrow{\psi^k - 1} & BSO^{\oplus}[p] \\
\downarrow{\scriptstyle \hat{\alpha}_p} & & \downarrow{\scriptstyle \alpha_p} & & \downarrow{\scriptstyle Id} \\
SG[p] & \longrightarrow & G/O[p] & \longrightarrow & BSO^{\oplus}[p] \\
\downarrow{\scriptstyle \hat{\sigma}_2} & & \downarrow{\scriptstyle \beta_p} & & \downarrow{\scriptstyle \bar{\rho}} \\
J_p^{\otimes} & \longrightarrow & BSO^{\otimes}[p] & \xrightarrow{1/\psi^k} & BSO^{\otimes}[p] .
\end{array}$$

The horizontal sequences are homotopy fiberings, and the maps $\beta_p \circ \alpha_p$ and $\bar{\rho}$ are homotopy equivalences. From the homotopy exact sequence it follows that $\hat{\sigma}_2 \circ \hat{\alpha}_p$ induces an isomorphism on each homotopy group. Thus $\hat{\sigma}_2 \circ \hat{\alpha}_p$ is a homotopy equivalence and

$$\alpha_p \times \text{incl}: J_p \times cokJ_p \to SG[p] \times SG[p] \to SG[p]$$

is the required homotopy equivalence.

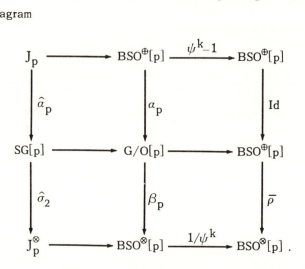

D. *The splitting of* MSPL[p], p *odd*

Suppose we are given a stable, piece-wise linear, bundle F over a CW complex X or equivalently a map $F: X \to BSPL$. Each skeleton $X^{(i)}$ of X is mapped into some finite $BSPL_{n(i)}$ so we get a PL-bundle $F_{n(i)}$ of fiber dimension n(i) over each $X^{(i)}$ and under restriction

$$F_{n(i)}|X^{(i-1)} \cong F_{n(i-1)} \oplus \varepsilon_i$$

with ε_i a trivial bundle. On the Thom space level we thus get maps $S^{m(i)} \wedge MF_{n(i-1)} \to MF_{n(i)}$, $m(i) = n(i) - n(i-1)$ and these define the Thom spectrum MF.

In 5.12 we constructed a splitting of BSPL at odd primes,

$$BSO^{\oplus}[p] \times BcokJ_p \xrightarrow{\gamma_p \times \text{incl.}} BSPL[p] \times BSPL[p] \xrightarrow{\oplus} BSPL[p] .$$

The Thom space of a product bundle is the smash product of the Thom spaces of the factors and similarly for Thom spectra. Hence we have an induced splitting

$$ME \wedge McokJ_p \simeq MSPL[p]$$

where E is represented by γ_p and where $McokJ_p$ is the Thom spectrum of the inclusion of $BcokJ_p$ in BSPL[p].

LEMMA 5.19. *The Thom spectrum* ME *is homotopy equivalent to* MSO[p].

Proof. The homotopy type of a Thom space depends only on the underlying spherical fibration. The PL-bundle E splits as a sum $E = E_1 \oplus E_2$ where E_1 sits over $BSO_{(1)}$ and E_2 sits over $BSO^{\perp}_{(1)}$. The restriction of γ_p to $BSO_{(1)}$ is the usual inclusion and the restriction of γ_p to $BSO^{\perp}_{(1)}$ is essentially the map $G/PL[p] \to BSPL[p]$. Hence E_2 is fiber homotopy trivial and E_1 is the restriction of the universal vector bundle over BSO[p]. On the other hand, since $\psi^k - 1: BSO^{\perp}_{(1)} \to BSO^{\perp}_{(1)}$ is a homotopy equivalence (by 5.8) the diagram

$$\begin{array}{ccc} & & G/O[p] \\ & \overset{\alpha_p}{\nearrow} & \downarrow \\ BSO^{\oplus}[p] & \xrightarrow{\psi^k - 1} & BSO^{\oplus}[p] \\ & & \downarrow \\ & & BSG[p] \end{array}$$

implies that the universal vector bundle ξ over $BSO^{\oplus}[p]$ is fiber homotopy trivial over the subspace $BSO^{\perp}_{(1)}$. This proves that E is fiber homotopically equivalent to ξ and the lemma follows.

COROLLARY 5.20 (Sullivan). *For any odd prime* p

$$MSPL[p] \simeq MSO[p] \wedge McokJ_p .$$

The space $cokJ_p$ is a direct factor in SG[p] and hence has finite homotopy and homology groups in each dimension. Thus rationally $MSO[p] \simeq MSPL[p]$.

But in fact more is true:

THEOREM 5.21. *For each odd prime*

$$\hat{\gamma}_p : \pi_*(MSO[p])/Tor \to \pi_*(MSPL[p])/Tor$$

is an isomorphism.

Our proof of 5.21 proceeds in a rather roundabout manner using the Hattori-Stong characterization of the lattice $\pi_*(MSO[p])/Tor$ in $H_*(MSO; \mathbf{Q})$ and we defer it to Chapter 11.B where we also prove a piecewise linear version of the Hattori-Stong theorem.

In 5.A we constructed a Thom class

$$\Delta_{PL} \in \widetilde{KO}(MSPL) \otimes \mathbf{Z}[1/2]$$

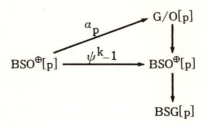

with Pontrjagin character $ph(\Delta_{PL}) = \mathcal{L}_{PL}^{-1} \cdot U_{PL}$ (cf. 5.4). Our procedure did not seem to characterize Δ_{PL} uniquely, and did not for example seem to allow proving the usual exponential property of Thom classes: $\Delta_{PL}(E\oplus F) = \Delta_{PL}(E)\cdot\Delta_{PL}(F)$. However, the next result due to Hodgkin for odd primes and to Hodgkin-Snaith in general implies that Δ_{PL} is in turn characterized by its Pontrjagin character.

THEOREM 5.22 (Hodgkin-Snaith). *For every prime* p, $\tilde{K}(\mathrm{cok}J_p) = 0$.

Proof. From the proof of 5.17 we have a fibration

$$\mathrm{cok}J_p \longrightarrow SG[p] \xrightarrow{\hat{\sigma}_2} J_p^{\otimes}$$

and from 5.18 $\hat{\sigma}_2$ is split:

$$J_p^{\oplus} \xrightarrow{a_p} SG[p] \xrightarrow{\hat{\sigma}_2} J_p^{\otimes}$$

is a homotopy equivalence. Hence it suffices to show that

$$\hat{\sigma}_2^*: \tilde{K}(J_p^{\otimes}) \to \tilde{K}(SG[p])$$

is a surjection. Now $J_p^{\otimes}$ is included in $BSO^{\otimes}[p]$ (as the homotopy fiber of ψ^k-1) and $BSO^{\otimes}[p]$ is included in $BU^{\otimes}[p]$ via complexification, so it is enough to calculate that

5.23 $$f = c \circ \mathrm{incl.} \circ \hat{\sigma}_2 : SG[p] \to BU^{\otimes}[p]$$

induces a surjection in K-theory. We outline this calculation in section 5.F below.

We have seen in 5.17 that $\mathrm{cok}J_p$ is a loop space for p odd,*) $\mathrm{cok}J_p = \Omega B\mathrm{cok}J_p$ and in particular we have the fibration

$$\mathrm{cok}J_p \to E\mathrm{cok}J_p \to B\mathrm{cok}J_p .$$

---

*) The space $\mathrm{cok}J_p$ is a loop space also for $p = 2$, in fact $\mathrm{cok}J_p$ is an infinite loop space at all primes. A proof of these somewhat more intricate facts can be found in [85].

There is a spectral sequence (the K-theoretic Eilenberg-Moore spectral sequence, see [56]) with E^2-term

$$E^2 = \mathrm{Tor}_{K_*(\mathrm{cok}J_p)}(K_*(pt)\otimes \mathbf{Z}_{(p)}, K_*(pt)\otimes \mathbf{Z}_{(p)})$$

$$E^\infty = K_*(B\mathrm{cok}J_p) .$$

Now, $K^{-1}(\mathrm{cok}J_p)$ is a direct factor of $K^{-1}(SG[p])$ but $K^{-1}(SG[p] = 0$ (cf. the proof of 5.17) so $\tilde{K}^*(\mathrm{cok}J_p) = 0$, and hence by the universal coefficient formula $\tilde{K}_*(\mathrm{cok}J_p) = 0$. The spectral sequence then gives $\tilde{K}_*(B\mathrm{cok}J_p) = 0$ and hence $\tilde{K}^*(B\mathrm{cok}J_p) = 0$. It follows that $\widetilde{KO}^*(B\mathrm{cok}J_p) = 0$ and using the Thom isomorphism we have (at odd primes)

COROLLARY 5.24. (i) $\widetilde{KO}^*(M\mathrm{cok}J_p) = 0$.
(ii) $\gamma_p^*: \widetilde{KO}(MSPL[p]) \to \widetilde{KO}(MSO[p])$ *is an isomorphism.*

In Chapter 4.E we proved that $\widetilde{KO}(BSO[p])$ is detected by the Pontrjagin character, and by the Thom isomorphism this is also true for $KO(MSO[p])$. Thus we have

COROLLARY 5.25. *The Pontrjagin character*

$$ph: \widetilde{KO}(MSPL[p]) \to H^*(MSPL; \mathbf{Q})$$

is injective (at odd primes).

We note as a consequence of 5.25 that the class $\Delta_{PL} \in \widetilde{KO}(MSPL[p])$ is uniquely characterized by 5.4 and that $\Delta_{PL}(E)$ is exponential for stable PL-bundles,

$$\Delta_{PL}(E\oplus F) = \Delta_{PL}(E)\cdot \Delta_{PL}(F) .$$

E. *Brumfiel's results*

For later use we recall Brumfiel's results on the homotopy groups in the diagram

5.26
$$\begin{array}{ccc} G/O & \xrightarrow{\ s\ } & G/PL \\ \Big\downarrow{\scriptstyle r} & & \Big\downarrow{\scriptstyle \hat{j}} \\ BSO & \xrightarrow{\ j\ } & BSPL\ . \end{array}$$

After making use of the Adams conjecture to give the order of the image of the J-homomorphism in dimensions $4n-1$ his results [31, p. 307] may be stated as follows: In dimension $4n$ the maps in homotopy of 5.26 may be summarized in

$$\begin{array}{ccc} Z \oplus \text{Torsion} & \xrightarrow{\ N_n\ } & Z \\ {\scriptstyle D_n}\Big\downarrow & & \Big\downarrow{\scriptstyle (\beta_n, \varepsilon_n, 0)} \\ Z & \xrightarrow{\ (\alpha_n, 1, 0)\ } & Z \oplus Z/2^{w(n)} \oplus \pi_{4n}(S^0)/\text{im}J \end{array}$$

for certain $\alpha_n, \beta_n, \varepsilon_n$ with

$$w(n) = \nu_2(n) + 3 \quad \text{for} \quad n > 2$$
$$w(1) = 0$$
$$w(2) = 2$$

and

$$N_n = a_n\, 2^{2n-2}(2^{2n-1}-1)\, \text{Num}(B_{2n}/4n)$$
$$D_n = \text{Denom}(B_{2n}/4n)\ .$$

Here $a_n = 1$ if n is even and $a_n = 2$ if n is odd, and B_n is the n'th Bernoulli number (cf. Chapter 11.A). At this point 5.26 is an easy exercise—which we leave for the reader to carry out—once we remark that the coefficient of p_{4n} in the Hirzebruch genus $\mathcal{L}$ is $8N_n/a_n(2n-1)!\,D_n$. In low dimensions one uses Cerf's result that PL/O is 6-connected.

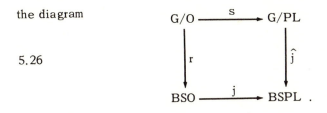

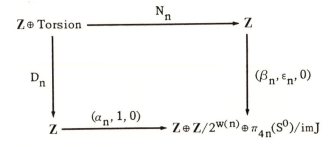

F. *The map* $f: SG[p] \to BU^{\otimes}[p]$

The proof of the Hodgkin-Snaith theorem (5.21) was based on the splitting of $SG[p]$ from 5.18 and the following calculational result

THEOREM 5.27. *The map* f *from 5.23 induces a surjection* $f^*: K(BU^{\otimes}[p]) \to K(SG[p])$.

The original proof of 5.27 used Hodgkins' calculation of $K_0(SG[p]; \mathbf{Z}/p^i)$ ([55], [56], [79]). The proof we outline below is based on a well-known reduction of 5.27 to a problem in representation theory which we then attack using transfer techniques. It is arithmetically very simple, but conceptually more involved.

Let $g_{2n}: S^{2n} \to BU$ represent the standard generator of $\tilde{K}(S^{2n})$. By Bott periodicity $\Omega^{2n}(BU) = BU \times \mathbf{Z}$, and the maps $\Omega^{2n} g_{2n}$ fit together to define a map

$$g: Q(S^0) \to BU \times \mathbf{Z},$$

so restricting to the 1-component $SG \subset Q(S^0)$ we obtain a map

$$g_1: SG \to BU \times (1) = BU^{\otimes}.$$

LEMMA 5.28. *The* p-*localization of* g_1 *is homotopic to* f.

Proof. Let X be a finite complex. A (based) map α from X_+ to the 1-component $\Omega_1^n S^n$ of $\Omega^n S^n$ can be thought of as a proper homotopy equivalence of the trivial bundle

$$\alpha: X \times \mathbf{R}^{n+1} \to X \times \mathbf{R}^{n+1}$$

(over Id_X), and the induced map of Thom spaces

$$M(\alpha): X_+ \wedge S^{n+1} \to X_+ \wedge S^{n+1}$$

is just the composition

$$X_+ \wedge S^{n+1} \xrightarrow{\text{diag} \wedge 1} X_+ \wedge (X_+ \wedge S^{n+1}) \xrightarrow{\text{id} \wedge \text{adj}(\Sigma\alpha)} X_+ \wedge S^{n+1}$$

where $\Sigma\alpha: X_+ \to \Omega^n S^n \subset \Omega^{n+1} S^{n+1}$ and $\text{adj}(\Sigma\alpha)$ denotes the adjoint map.

We assume that $n+1$ is even. Then the Thom class $\Delta_U \in K(X_+ \wedge S^{n+1})$ is $1 \otimes g_{n+1}$, and $[f \circ \alpha] \in K(X)$ is the element satisfying

$$[f \circ \alpha] \cdot \Delta_{SO} = M(\alpha)^*(\Delta_{SO}) .$$

Now it is direct that $f \circ \alpha$ and $g \circ (\Sigma\alpha)$ represents the same element of $K(X)$, so passing to the limit over n the result follows.

Both the domain and the range of g are infinite loop spaces and by construction g is an infinite loop map. (See Chapter 6 for a brief discussion of infinite loop spaces and [50], [84] and [85] for further details.) Each infinite loop space E admits a sequence of structure maps

$$d_n : E\Sigma_n \times_{\Sigma_n} E^n \to E$$

and these are natural w.r.t. infinite loop maps. In particular we have a homotopy commutative diagram

5.29
$$\begin{array}{ccc} E\Sigma_n \times_{\Sigma_n} Q(S^0)^n & \xrightarrow{d_n} & Q(S^0) \\ \Big\downarrow {\scriptstyle 1 \times_{\Sigma_n} g^n} & & \Big\downarrow {\scriptstyle g} \\ E\Sigma_n \times_{\Sigma_n} (BU \times Z)^n & \xrightarrow{d_n^U} & BU \times Z . \end{array}$$

The structure map d_n^U in 5.29 has a simple bundle theoretical description. Let ζ_r denote the universal r-dimensional bundle over BU_r.

LEMMA 5.30. *The restriction of* d_n^U *to* $E\Sigma_n \times_{\Sigma_n} (BU_r)^n$ *represents the bundle* $E\Sigma_n \times_{\Sigma_n} \zeta_r \oplus \cdots \oplus \zeta_r$ *(of dimension* rn*). (See e.g.* [85] *for a proof.)*

Theorem 3.9 asserts the existence of certain natural inclusions $i_n : B\Sigma_n \to Q(S^0)$. They can be explicitly obtained from the structure maps d_n via the composition

$$i_n : B\Sigma_n \xrightarrow{\simeq} E\Sigma_n \times_{\Sigma_n} (1) \times \cdots \times (1) \longrightarrow E\Sigma_n \times_{\Sigma_n} Q(S^0)^n \xrightarrow{d_n} Q(S^0)$$

where $(1) \subset Q(S^0)$ is the map representing the identity on the stable sphere. From 5.29 and 5.30 we have

COROLLARY 5.31. *Let* $P : \Sigma_n \to U_n$ *be the permutation representation. The diagram*

$$\begin{array}{ccc} B\Sigma_n & \xrightarrow{BP} & BU_n \\ \Big\downarrow{\scriptstyle i_n} & & \Big\downarrow \\ Q_n(S^0) & \xrightarrow{g} & BU \times (n) \subset BU \times \mathbf{Z} \end{array}$$

is homotopy commutative.

The inclusion $P : \Sigma_n \to U_n$ induces a homomorphism of the complex representation rings

$$P^* : R(U_n) \to R(\Sigma_n) .$$

Then to prove 5.27 we prove

THEOREM 5.32. $P^* : R(U_n) \to R(\Sigma_n)$ *is surjective.*

It is classical that the representation ring $R(\Sigma_n)$ is additively generated by the trivial 1-dimensional representation and representations induced up from the subgroups $\Sigma_{i_1} \times \cdots \times \Sigma_{i_r} \subset \Sigma_n$, $(i_1 + \cdots + i_r = n)$. We refer the reader to [154] for a very elegant proof of this. Thus to prove 5.32 it suffices to construct induction homomorphisms

$$R(U_n) \otimes R(U_m) = R(U_n \times U_m) \xrightarrow{(i_{n,m})_!} R(U_{n+m})$$

commuting with the classical induction homomorphisms associated to $\Sigma_n \times \Sigma_m \subset \Sigma_{n+m}$.

We use a construction due to Becker-Gottlieb, the continuous transfer ([18], [34]). It is a stable map

$$\tau : BG^+ \to BH^+$$

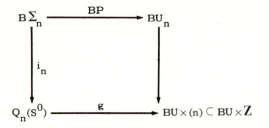

where H is a closed subgroup of the compact Lie group G.*) Alternatively, τ is a map $BG \to Q(BH^+)$.

If G and H are finite groups then the induced map

$$\tau^* : H^*(BH; \mathbf{Z}) \to H^*(BG; \mathbf{Z})$$

agrees with the classical transfer considered in Chapter 3.C. More generally though, for any cohomology theory E^*, τ induces a map

$$\tau^* : E^*(BH) \to E^*(BG) .$$

The composition $BG \xrightarrow{\tau} Q(BH^+) \xrightarrow{c} Q(S^0)$ (where c collapses BH to a point) defines a stable cohomotopy class $\chi(H:G) \in \pi_S^0(BG)$. Now, each generalized cohomology theory is a module over stable cohomotopy, so we have an action of $\pi_S^0(BG)$ on $E^*(BG)$. Becker and Gottlieb have given the following important generalization of Lemma 3.13.

THEOREM 5.33 (Becker-Gottlieb). (i) *Let* $\pi^* : E^*(BG) \to E^*(BH)$ *be induced from the inclusion* $H \subset G$. *Then*

$$\tau^* \pi^*(x) = \chi(H:G) \cdot x$$

(ii) *The degree of* $\chi(H:G)$ *is the Euler characteristic of the fiber* G/H.

Next, we specialize to G connected and $H \subset G$ a subgroup of maximal rank. Let $T \subset H$ be a maximal torus and let W_G, W_H be the Weyl groups. Then W_G acts (from the right) on BT; let $\sigma \cdot x$ denote the induced left action on $x \in E^*(BT)$ and note from Lemma 3.11 that $\sigma \cdot x$ depends only on the coset of σ in W_G/W_H when x is the restriction of an element from $E^*(BH)$. From [34], p. 142 we have

THEOREM 5.34. *For* $x \in E^*(BH)$

$$j_G^* \circ \tau^*(x) = \sum \sigma \cdot j_H^*(x)$$

where the summation extends over W_G/W_H *and* j_H, j_G *denote the inclusions of* BT *in* BH *and* BG.

---

*)More generally, the continuous transfer $\tau : X^+ \to Y^+$ is defined for every fibering $Y \to X$ with compact fiber F.

We shall use this result only when E^* is complex K-theory. Recall from Chapter 4.E that $K(BG)$ is the completion of the complex representation ring $R(G)$. Since G is connected $R(G)$ injects into its completion. Moreover, $R(G)$ injects into $R(T)$ as the subgroup invariant under the Weyl group, so we can use $R(T)$ to name elements of both $R(H)$ and $R(G)$. It follows that

$$\tau^* : K(BH) \to K(BG)$$

maps $R(H)$ to $R(G)$, and we have

COROLLARY 5.35. *There exist induction homomorphisms*

$$\mathrm{Ind}_G^H : R(H) \to R(G)$$

with

$$\mathrm{Ind}_G^H(x) = \Sigma\, \sigma \cdot x, \qquad \sigma \in W_G/W_H .$$

We now specialize to the unitary groups $H = U_n \times U_m$, $G = U_{n+m}$ and $T = U_1^{n+m}$, embedded as the set of diagonal matrices.

COROLLARY 5.36. *There are induction homomorphisms*

$$\mathrm{Ind} : R(U_n \times U_m) \to R(U_{n+m})$$

so that the diagrams

$$\begin{array}{ccc} R(U_n \times U_m) & \xrightarrow{\ \mathrm{Ind}\ } & R(U_{n+m}) \\ \Big\downarrow{\scriptstyle (P\times P)^*} & & \Big\downarrow{\scriptstyle P^*} \\ R(\Sigma_n \times \Sigma_m) & \xrightarrow{\ \mathrm{Ind}\ } & R(\Sigma_{n+m}) \end{array}$$

are commutative.

Proof. Let $N(n, m)$ be the normalizer of U_1^{n+m} in $U_n \times U_m$ and $N(n+m)$ the normalizer in U_{n+m}. The Weyl groups of $U_n \times U_m$ and U_{n+m} are $\Sigma_n \times \Sigma_m$ and

Σ_{n+m} respectively, and they are naturally embedded in the normalizers. Let V be a representation of $H = U_n \times U_m$. The restriction of the induced representation $\mathrm{Ind}(V)$ to the normalizer $N(n+m)$ is given by $N(n+m) \times_{N(n,m)} V$; this follows e.g. by restricting to the torus and using 5.35. But then it is direct to see that

$$P^*(\mathrm{Ind}(V)) = \Sigma_{n+m} \times_{\Sigma_n \times \Sigma_m} V$$

where $\Sigma_n \times \Sigma_m$ acts on V via the inclusion $P \times P$. This proves 5.36.

Finally, from 5.36 we see that $R(U_n) \xrightarrow{P^*} R(\Sigma_n)$ is onto (compare the paragraph following 5.32) and then by completion that $K(BU_n) \to K(B\Sigma_n)$ is surjective.*) Now 5.27 follows from 3.11, 5.28 and 5.31.

---

*) Let $K = \mathrm{Ker}\{R(U_n) \to R(\Sigma_n)\}$. Since $\{K/K \cap I(U_n)^\ell\}_\ell$ is an inverse system with vanishing $\varprojlim^{(1)}$ the claimed surjectivity follows from 4.23.

CHAPTER 6

INFINITE LOOP SPACES AND THEIR HOMOLOGY OPERATIONS

The classifying spaces for the stable bundle theories, BO, BTOP, BPL and BG, and the "homogeneous" spaces G/TOP, G/O etc., are all infinite loop spaces and the natural maps between them are infinite loop maps by theorems of Boardman and Vogt [19]. We have already made several general remarks on the structure of infinite loop spaces in Chapter 3.B and in Chapter 5.F. However, for later results we need some further properties of such spaces. In particular, we require more information on the homology structure of the infinite loop space SG and its classifying space BSG.

The homology of an infinite loop space X admits operations

$$Q^a : H_i(X) \to H_{i+a}(X) ,$$

natural with respect to infinite loop maps, and extending the square in the Pontrjagin ring.*) The definition of Q^a is quite analogous to the definition of the Steenrod operation

$$Sq^a : H^i(X) \to H^{i+a}(X) ,$$

which exists for any space X. Indeed, the Steenrod operations measure the deviation from strict commutativity of the cochain level cup product, and similarly the homology operations measure the deviation from commutativity of the H-space multiplication. More precisely, the multiplication in an infinite loop space is homotopy commutative by a homotopy which is itself homotopy commutative and so forth, and one gets an infinite sequence of higher homotopies. The homology operations provide a measure of the non-trivialy of these homotopies.

Standard formulae for Steenrod operations (e.g. the Cartan formula and the Adem relations) are consequences of certain (homotopy) commutative diagrams on the cochain level (see e.g. [46], [130]). The existence of the relevant diagrams

---

*)Homology operations were originally discovered by Araki and Kudo [7]. The custom, quite inexplicably, is to call them Dyer-Lashof operations.

follows from the technique of acyclic models. There are similar diagrams (on the space or chain level) for infinite loop spaces, giving analogous formulae, but here one has to rely on the geometry of infinite loop spaces instead of acyclic models.

In the three volumes [84], [85] and [86], J. P. May and his collaborators have given a very thorough account of infinite loop space theory. We shall only need a small part of this theory, and the present chapter should serve the reader as a guide through the part of the theory relevant to this book.

Throughout the chapter $H_*(X)$ will denote homology with $\mathbf{Z}/2$ coefficients.

A. *Homology operations*

Suppose X is an n'th loop space, $X = \Omega^n Y$. Then there is a natural map

$$u : \Omega^n S^n X \to X$$

with $u \circ i = \mathrm{Id}$ where $i : X \to \Omega^n S^n X$ is the natural inclusion. Indeed, if $v : S^n X \to Y$ is the evaluation map, then $u = \Omega^n(v)$.

Formally, an infinite loop space X is a space together with a distinct sequence of "deloopings"

$$BX, B^2X, \cdots, B^iX, \cdots$$

$(\Omega^n B^n X \simeq X)$, that is, X is a single space in a (connected) Ω-spectrum. For such a space the maps u above fit together to give a retraction

$$u : Q(X) \to X$$

of the embedding $i : X \to Q(X)$, where as usual

$$Q(X) = \lim_{\to} \Omega^n S^n(X) .$$

Let $\Gamma_n(X) = S^{n-1} \times_{\Sigma_2} X \times X$ be the orbit space of the free Σ_2 action on $S^{n-1} \times X \times X$ which is antipodal on S^{n-1} and interchanges factors on $X \times X$. There is a natural map

6.1 $$D_2 : \Gamma_n(X) \to \Omega^n S^n(X), \qquad n \leq \infty$$

satisfying $D_2(1; x,y) = i(x) * i(y)$, where $*$ is the loop sum in $\Omega^n S^n(X)$ (cf. [28], [50]).

We are mainly interested in the case $n = \infty$. Here the existence of D_2 follows from the Dyer-Lashof "model" $C(X)$ for $Q(X)$, reviewed in Chapter 3.B. Indeed, D_2 is the composition

$$D_2 : \Gamma_\infty(X) \xrightarrow{j} C(X) \xrightarrow{i_X} Q(X) .$$

The first embedding j comes from the identification $S^\infty = E\Sigma_2$ and the second map i_X is a natural homotopy equivalence when X is connected, and a "group completion" map in general: $(i_X)_* : \pi_0(C(X)) \to \pi_0(Q(X))$ is an algebraic group completion and $(i_X)_* : H_*(C(X)) \to H_*(Q(X))$ localizes $H_*(C(X))$ at $\pi_0(C(X))$.

REMARK 6.2. There is an alternative way of seeing D_2, probably first noticed by Boardman, which has played an important role in the development of infinite loop spaces (see in particular [84]). The space $\Omega^n S^n(X)$ consists of all maps $x : (D^n, \partial D^n) \to (S^n X, *)$. If $x,y \in X$ we define $D_2(w; x,y)$ according to the picture

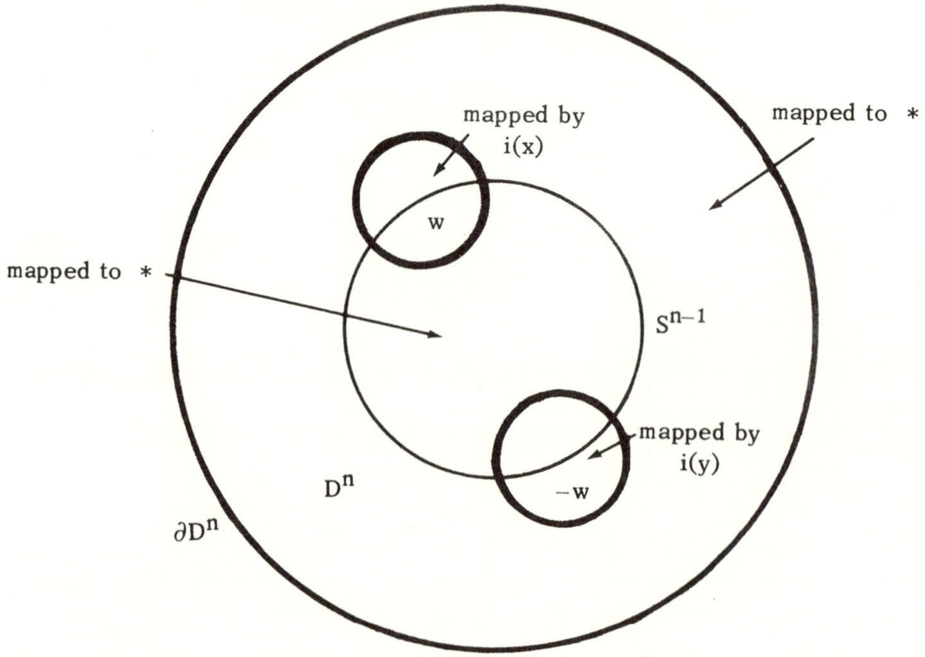

The two small discs around w and $-w$ are mapped according to $i(x)$ and $i(y)$ and their complement is mapped to the basepoint $*$.

Let X be an infinite loop space. We can then compose D_2 with the retraction u to obtain structure maps ("higher homotopies")

6.3 $$d_2 = u \circ D_2 : E\Sigma_2 \times_{\Sigma_2} X \times X \to X .$$

The maps D_2 are natural in X (since i_X is) and the retraction maps u_X are natural for maps of infinite loop spaces. Hence the structure maps d_2 are natural for infinite loop maps.*)

Using the notation of 3.20 we make the following basic definition ([7], [28], [50]).

DEFINITION 6.4. Let X be an infinite loop space. The j'th (lower) Dyer-Lashof operation

$$Q_j : H_i(X) \to H_{j+2i}(X)$$

is defined by

$$Q_j(a) = (d_2)_*(e_j \otimes a \otimes a)$$

where $a \in H_i(X)$.

REMARK 6.5. If X is just an n fold loop space, then the same definition applies to define $Q_j(a)$ for $j < n$. We only need to note that $e_j \otimes a \otimes a \in H_{j+2i}(\Gamma_n(X))$ if $j < n$.

Readers familiar with cohomology operations will notice the formal similarity between 6.4 and the definition of the (mod 2) Steenrod squares [46], [130]. It is not surprising therefore that the Q_j share many formal properties with the Sq^j.

*)An infinite loop map $f: X \to Y$ is a map which is the restriction of a homomorphism of the associated Ω-spectra.

THEOREM 6.6. (i) *If* $f: X \to Y$ *is an infinite loop map, then*

$$Q_j(f_*(\alpha)) = f_*(Q_j(\alpha))$$

for all $x \in H_*(X)$.

(ii) $Q_0(\alpha) = \alpha * \alpha$, *where* $*$ *denotes the Pontrjagin product in* $H_*(X)$.

(iii) *Let* $e \in H_0(X)$ *be the unit element in the Pontrjagin ring. Then* $Q_j(e) = 0$ *for* $j > 0$.

(iv) *Let* X *be an infinite loop space and* $v: S\Omega X \to X$ *the evaluation map. Then*

$$v_*(\Sigma Q_j(\alpha)) = Q_{j-1}(v_*(\Sigma\alpha))$$

where $\Sigma: H_*(\Omega(X)) \to H_*(S\Omega(X))$ *denotes the suspension homomorphism.*

(v) *Let* $\Delta: X \to X \times X$ *be the diagonal and let* $\Delta_*(\alpha) = \Sigma \alpha'_k \otimes \alpha''_k$. *Then*

$$\Delta_* Q_j(\alpha) = \Sigma Q_{j-r}(\alpha'_k) \otimes Q_r(\alpha''_k)$$

(vi) $Q_j(\alpha * \beta) = \Sigma Q_{j-r}(\alpha) * Q_r(\beta)$ *for all* $\alpha, \beta \in H_*(X)$.

(The first two properties are direct from the definitions and (iii) follows since $E\Sigma_2 \times_{\Sigma_2} (*) \times (*)$ is identified with the base point in $C(X)$. The last three properties are proved by evaluating certain homotopy commutative diagrams in homology. The relevant diagrams can be found in [84 Lemma 5.6, Lemma 5.7 and Lemma 1.9 (ii) (with $\mathcal{C}_n(2)$ replaced by S^{n-1})] and the homological calculations are given in [86, p. 7-9]. Alternatively, the reader might consult the original source [50].)

The next result is analogous to the Adem relations for Steenrod squares. It expresses a composition $Q_i Q_j(\alpha)$ with $i > j$ in terms of compositions with $i \leq j$. The precise formula is most conveniently described in the notation of "upper" Dyer-Lashof operations, Q^i. The connection is as follows:

6.7 $$Q^k(\alpha) = Q_{k-|\alpha|}(\alpha), \qquad |\alpha| = \deg(\alpha).$$

Thus Q^k raises degree by k. In the rest of the chapter we shall pass freely between "upper" and "lower" notation.

We denote by $\binom{a}{b}$ the binomial coefficient $a!/b!(a-b)!$ (mod 2) with the usual convention that $\binom{a}{b} = 0$ unless $a \geq b \geq 0$.

THEOREM 6.8. *For* $k > 2\ell$,

$$Q^k Q^\ell(a) = \sum_{2\nu \geq k} \binom{\nu-\ell-1}{2\nu-k} Q^{k+\ell-\nu} Q^\nu(a)$$

for all $a \in H_*(X)$.

(The proof of 6.8 can be found in [86, p. 9] (see also [50] and [130, p. 117].) It is completely analogous to the proof of the "classical" Adem relations once one has the commutative diagram

6.9

$$\begin{array}{ccccc} \Gamma_\infty(\Gamma_\infty(X)) & & \longrightarrow & & E(\Sigma_2 \wr \Sigma_2) \times_{\Sigma_2 \wr \Sigma_2} X^4 \\ \Big\downarrow \Gamma_\infty(d_2) & & & & \Big\downarrow \\ \Gamma_\infty(X) & \xrightarrow{d_2} & X & \xleftarrow{d_4} & E(\Sigma_4) \times_{\Sigma_4} X^4 \end{array}$$

Here d_4 is defined in analogy with d_2 above as the composition

$$d_4 : E\Sigma_4 \times_{\Sigma_4} X^4 \longrightarrow C(X) \xrightarrow{i_X} Q(X) \xrightarrow{u} (X) .$$

Finally, we have a formula for the interaction of Steenrod operations and homology operations first described by Nishida in [108]. Let

$$Sq_*^j : H_i(X) \to H_{i-j}(X)$$

be the vector space dual of the Sq^j acting on cohomology so that $\langle Sq^j(\xi), a \rangle = \langle \xi, Sq_*^j(a) \rangle$, where $\langle\ ,\ \rangle$ denotes the usual Kronecker pairing. From 3.21 we have (after dualizing)

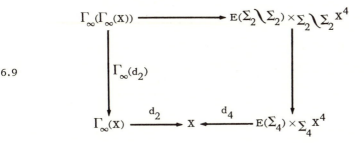

6.10 $$(1\times\Delta)_*(e_i \otimes \alpha) = \Sigma\, e_{i+2k-|\alpha|} \otimes Sq_*^k(\alpha) \otimes Sq_*^k(\alpha)$$

where $|\alpha| = \deg \alpha$ and $1\times\Delta : RP^\infty \times X \to \Gamma_\infty(X)$.

THEOREM 6.11. $Sq_*^i Q^j(\alpha) = \sum_{2k\leq i} \binom{j-i}{i-2k} Q^{j-i+k}(Sq_*^k(\alpha))$.

(The proof of 6.11 is a consequence of 6.10 and the classical Adem relation for Steenrod squares. It also involves some binomial arithmetic: we refer the reader to [108].)

B. *Homology operations in* $H_*(Q(S^0))$ *and* $H_*(SG)$

The calculations in Chapter 3 of the Pontrjagin rings $(H_*(Q(S^0)), *)$ and $(H_*(SG), \circ)$ were based primarily on two facts: the relations of the two products to cartesian product and disjoint sum of permutations (cf. 3.9 and 3.10), and the distributivity of the composition product $(\circ)$ over the loop sum $(*)$ (cf. Chapter 3.F).

Both $Q(S^0)$ and SG are infinite loop spaces (with classifying spaces $Q(S^1)$ and BSG, respectively) and we now wish to extend the considerations of Chapter 3.B and 3.F to also get information on the two (distinct) sets of homology operations.

To avoid confusion we denote the structure map for $Q(S^0)$ by d_2 and the structure map for SG by $\hat{d}_2$,

$$d_2 : E\Sigma_2 \times_{\Sigma_2} Q(S^0) \times Q(S^0) \to Q(S^0)$$

$$\hat{d}_2 : E\Sigma_2 \times_{\Sigma_2} SG \times SG \to SG .$$

Then $d_2(1; x,y) = x * y$ and $\hat{d}_2(1; x,y) = x \circ y$. The first map d_2 was made explicit in 6.2 above and for the second map $\hat{d}_2$ we have (cf. [73, p. 239] or [85, p. 9-18]).

REMARK 6.12. Let R^∞ be the direct limit of the R^n, consisting of all sequences $(x_1, x_2, \cdots)$ with almost all entries equal to zero. This is an inner product space with contractible isometry group $O(R^\infty, R^\infty)$ (Boardman, see [84] for a proof). Now $R^\infty \oplus R^\infty \cong R^\infty$ and $O(R^\infty \oplus R^\infty, R^\infty)$ is a Σ_2-space: $(f \cdot T)(x,y) = f(y,x)$. Let

$$\rho : S^\infty \to O(R^\infty \oplus R^\infty, R^\infty)$$

be a Σ_2-equivariant map, $\rho(-w) = \rho(w)T$ (see e.g. [73, p. 239] for a specific choice of ρ). The one point compactification of R^∞ is S^∞ and the structure map $\hat{d}_2$ is explicitly given by

$$\hat{d}_2(w; x,y) = \rho(w)(x \wedge y)\rho(w)^{-1}$$

where $\wedge$ denotes the smash product. The right hand side only affects finitely many coordinates of $(R^\infty)^+ = S^\infty$, and hence represents an element of SG.

Although $Q(S^0)$ is not an infinite loop space in the composition structure $((\pi_0(QS^0), \circ) = (Z, \cdot)$ is not a group) it follows from 6.12 that the structure map $\hat{d}_2$ has a natural extension to a map

$$\hat{d}_2 : E\Sigma_2 \times_{\Sigma_2} Q(S^0) \times Q(S^0) \to Q(S^0) .$$

Note that the extension $\hat{d}_2$ satisfies

$$\hat{d}_2(w; x,0) = \hat{d}_2(w; 0,x) = 0$$

where 0 denotes the constant map (the base point of $Q_0(S^0)$).

LEMMA 6.13. *The two maps* f, g *below from* $E\Sigma_2 \times_{\Sigma_2} Q(S^0)^2 \times Q(S^0)^2$ *to* $Q(S^0)$ *are homotopic:*

$$f(w; (x_1,y_1), (x_2,y_2)) = \hat{d}_2(w; x_1 * y_1, x_2 * y_2)$$

$$g(w; (x_1,y_1), (x_2,y_2)) = \hat{d}_2(w; x_1,x_2) * \hat{d}_2(w; y_1,y_2) * d_2(w; x_1 \wedge y_2, x_2 \wedge y_1) .$$

(The proof of 6.13 consists of a careful study of

$$f(w;(x_1,y_1),(x_2,y_2)) = \rho(w)((x_1 * y_1)\wedge(x_2 * y_2))\rho(w)^{-1}.$$

Indeed, from the distributivity of $\wedge$ over $*$ we get directly a decomposition of the form listed for g except that the third term is $\rho(w)((x_1\wedge y_2)*(y_1\wedge x_2))\rho(w)^{-1}$ instead of $d_2(w;x_1\wedge y_2,x_2\wedge y_1)$. However, using the explicit description of d_2 from 6.2 one can check that the two expressions define homotopic maps from $E\Sigma_2\times_{\Sigma_2}Q(S^0)^2\times Q(S^0)^2$ to $Q(S^0)$. We leave the details to the reader with the warning that the argument is not quite trivial: it represents a real shortcut to the "mixed Cartan formula" 6.15 below, cf. [73] and [84]).

The structure maps $d_2, \hat{d}_2$ give two sets of Dyer-Lashof operations

$$Q_j, \hat{Q}_j : H_i(Q(S^0)) \to H_{j+2i}(Q(S^0)).$$

Here Q_j is the operation associated with the infinite loop space structure on $Q(S^0)$ $(B^n(Q(S^0)) = Q(S^n))$ and the restriction of $\hat{Q}_j$ to $H_*(SG)$ is the operation associated to the Boardman-Vogt infinite loop space structure on SG. For each $\alpha \in H_*(Q(S^0))$,

$$Q_0(\alpha) = \alpha * \alpha. \quad \hat{Q}_0(\alpha) = \alpha \circ \alpha$$

and, in general, if $\alpha \in H_*(Q_k(S^0))$ then

$$Q_j(\alpha) \in H_*(Q_{2k}(S^0)), \quad \hat{Q}_j(\alpha) \in H_*(Q_{k^2}(S^0))$$

where $Q_k(S^0)$ is the component of $Q(S^0)$ consisting of maps of degree k.

If $[k] \in H_0(Q_k(S^0))$ is the non-zero element, then for $j > 0$ we have

6.14 $$Q_j([0]) = 0, \quad \hat{Q}_j([1]) = 0 \quad \text{and} \quad \hat{Q}_j([0]) = 0$$

but neither set of operations is trivial on zero-dimensional elements in general.

THEOREM 6.15. *Let $\alpha, \beta \in H_*(Q(S^0))$ be arbitrary classes with coproducts*

$$\psi(\alpha) = \sum \alpha'_k \otimes \alpha''_k, \quad \psi(\beta) = \sum \beta'_\ell \otimes \beta''_\ell .$$

Then

$$\hat{Q}^i(\alpha * \beta) = \sum \hat{Q}^{i_1}(\alpha'_k) * \hat{Q}^{i_2}(\beta'_\ell) * Q^{i_3}(\alpha''_k \circ \beta''_\ell)$$

where the summation extends over all pairs k, ℓ *and triples* (i_1, i_2, i_3) *with* $i = i_1 + i_2 + i_3$.

(The proof is direct from 6.13. Consider the element

$$A = e_{i-|\alpha|-|\beta|} \otimes (\alpha \otimes \beta) \otimes (\alpha \otimes \beta)$$

in $H_*(E\Sigma_2 \times_{\Sigma_2} Q(S^0)^2 \times Q(S^0)^2)$. We clearly have $f_*(A) = \hat{Q}^i(\alpha * \beta)$, and $g_*(A)$ is indeed the right hand side of the required formula. The last assertion uses the standard diagonal approximation

$$\psi : C_*(S^\infty) \to C_*(S^\infty) \otimes C_*(S^\infty)$$

given explicitly by $\psi(e_n) = \Sigma e_i \otimes e_{n-i} T^i$ (where $e_n \in C_n(S^n) \subset C_n(S^\infty)$ is represented by the upper half disc). It also requires the fact that smash product and composition product induce homotopic H-structures in SG. The reader might check [73, p. 248-250] or [84, p. 84-92] for an alternative proof.)

We now wish to relate $d_2, \hat{d}_2$ to embeddings

$$J_{2,m} : \Sigma_2 \backslash \Sigma_m \to \Sigma_{2m}$$

$$\hat{J}_{2,m} : \Sigma_2 \backslash \Sigma_m \to \Sigma_{m^2} .$$

The first of these was defined in 3.15 and $\hat{J}_{2,m}$ is a multiplicative analogue: consider Σ_{m^2} as the automorphism group of pairs (i, j); then the embedding $\hat{J} = \hat{J}_{2,m}$ is specified by

$$\hat{J}(1; h_1, h_2)(i, j) = (h_1(i), h_2(j))$$

$$\hat{J}(\sigma; 1, 1)(i, j) = (j, i) .$$

Let $i_m : B\Sigma_m \to Q(S^0)$ be the map given in 3.9. We can strengthen 3.9 and 3.10 to (see [85, Chapter 6]).

THEOREM 6.16. *The diagrams below homotopy commute for all* m

$$\begin{array}{ccc} \Gamma_\infty(B\Sigma_m) & \xrightarrow{\Gamma_\infty(i_m)} & \Gamma_\infty(Q(S^0)) \\ \Big\downarrow{\scriptstyle BJ_{2,m}} & & \Big\downarrow{\scriptstyle d_2} \\ B\Sigma_{2m} & \xrightarrow{i_{2m}} & Q(S^0) \end{array} \qquad \begin{array}{ccc} \Gamma_\infty(B\Sigma_m) & \xrightarrow{\Gamma_\infty(i_m)} & \Gamma_\infty(Q(S^0)) \\ \Big\downarrow{\scriptstyle B\hat{J}_{2,m}} & & \Big\downarrow{\scriptstyle \hat{d}_2} \\ B\Sigma_{m^2} & \xrightarrow{i_{m^2}} & Q(S^0) \end{array}$$

Here we have identified $\Gamma_\infty(B\Sigma_m)$ *with* $B(\Sigma_2 \wr \Sigma_m)$ (cf. 3.16).

REMARK 6.17. We have two conjugate embeddings of $\Sigma_2 \wr (\Sigma_n \times \Sigma_m)$ in $\Sigma_{(n+m)^2}$

$$\phi : \Sigma_2 \wr (\Sigma_n \times \Sigma_m) \xrightarrow{\Sigma_2 \wr *} \Sigma_2 \wr \Sigma_{n+m} \xrightarrow{\hat{J}} \Sigma_{(n+m)^2} .$$

$$\psi : \Sigma_2 \wr (\Sigma_n \times \Sigma_m) \xrightarrow{\tau} \Sigma_2 \wr \Sigma_n \times \Sigma_2 \wr \Sigma_m \times \Sigma_2 \wr \Sigma_{nm} \xrightarrow{\hat{J}\times\hat{J}\times J} \Sigma_{n^2} \times \Sigma_{m^2} \times \Sigma_{2nm} \xrightarrow{*} \Sigma_{(n+m)}$$

where

$$\tau(\sigma; (h_1,g_1), (h_2,g_2)) = ((\sigma; h_1,h_2), (\sigma,g_1,g_2), (\sigma; \psi_{n,m}(h_1,g_2)\psi_{n,m}(h_2,g_1))$$

(cf. 3.10). Then $B\phi \simeq B\psi$, and from 6.16 it follows that $B\phi$ and $B\psi$ are the restrictions of the maps f and g from 6.13 to the subspace $E\Sigma_2 \times_{\Sigma_2} (B\Sigma_n \times B\Sigma_m) \times (B\Sigma_n \times B\Sigma_m)$ of $E\Sigma_2 \times_{\Sigma_2} Q(S^0)^2 \times Q(S^0)^2$. In view of 3.9 this strongly suggests 6.13 but it does not quite seem to give an alternative argument.

Our next result is a specialization of a result due to May (see [84, p. 11] or [73, p. 247]).

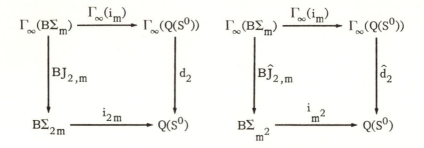

THEOREM 6.18. *For each* $a \in H_*(Q(S^0))$

$$Q^i[1] \circ a = \sum Q^{i+k}(Sq_*^k(a)) .$$

Proof. From 3.9 and 3.40 (ii) it suffices to prove the result when a belongs to the image of $(i_m)_* : H_*(B\Sigma_m) \to H_*(Q(S^0))$ for some m. But $J_{2,m} \circ (1 \times \Delta) = \psi_{2,m}$ where $1 \times \Delta : \Sigma_2 \times \Sigma_m \to \Sigma_2 \backslash \Sigma_m$ and $\psi_{2,m}$ is the cartesian product (cf. p. 53); the formula follows from 6.10.

We shall see in Chapter 6.C below that the elements $Q^i[1] \in H_i(Q(S^0))$ and $[-1] \in H_0(Q(S^0))$ generate $H_*(Q(S^0))$ under application of both the composition and the loop product. From 6.6 (vi), 6.14, 6.15 and 6.18 we have

$$0 = \hat{Q}^i([-1] * [1]) = \sum_{j=0}^{i} \hat{Q}^j[-1] * Q^{i-j}[-1] * [1]$$

$$0 = Q^i([-1] * [1]) = \sum_{j=0}^{i} Q^j[-1] * Q^{i-j}[1] .$$

That is to say,

$$\hat{Q}^i([-1]) * [1] = \chi(Q^i[-1])$$

$$Q^i[-1] = \chi(Q^i[1])$$

where $\chi : H_*(Q(S^0)) \to H_*(Q(S^0))$ is the canonical (anti)automorphism of the Hopf algebra $(H_*(Q(S^0)), *)$ (cf. 3.43). Thus we have

6.19 $$\hat{Q}^i([-1]) = Q^i[1] * [-1] .$$

In the proof of the next result we need a simple summation formula for mod 2 binomial coefficients,

6.20 $$\sum_{\ell=0}^{c} \binom{b+\ell}{\ell}\binom{a}{c-\ell} = \binom{a-b-1}{c} \text{ for } a > b$$

$$= \binom{b+c-a}{c} \text{ for } a \leq b .$$

(To prove this let $[{x \atop i}]$ be the coefficient of t^i in $(1+t)^x \in Z/2[[t]]$. Then $({a \atop b}) = [{b-a-1 \atop b}]$ whenever $a \geq b \geq 0$, so $({b+\ell \atop \ell}) = [{-b-1 \atop \ell}]$. The polynomial identity $(1+t)^{-b-1}(1+t)^a = (1+t)^{a-b-1}$ now gives the formula.)

THEOREM 6.21. *$\hat{Q}^i(Q^j[1]) = \sum_{k=0}^{j} ({i-k-1 \atop j-k})\, Q^k[1] * Q^{i+j-k}[1]$, $i > j$ and $\hat{Q}^i(Q^i[1]) = Q^i[1] * Q^i[1]$.*

Proof. Consider the two conjugate embeddings

$$s: \Sigma_2 \times \Sigma_2 \xrightarrow{1\times\Delta} \Sigma_2 \wr \Sigma_2 \xrightarrow{\hat{J}} \Sigma_4$$

$$t: \Sigma_2 \times \Sigma_2 \xrightarrow{1\times\Delta} \Sigma_2 \times \Sigma_2 \times \Sigma_2 \xrightarrow{\mu\times 1} \Sigma_2 \times \Sigma_2 \xrightarrow{*} \Sigma_4$$

where $\mu: \Sigma_2 \times \Sigma_2 \to \Sigma_2$ is the multiplication in Σ_2.

Let $e_a \in H_*(B\Sigma_2)$ be the unique non-zero element of degree a. From 6.16 (with $m = 1$) we have $(i_2)_*(e_a) = Q^a[1] \in H_a(Q(S^0))$, and from 6.10 and 6.16 we get

$$(i_4)_*(Bs)_*(e_i \otimes e_j) = \sum ({j-k \atop k})\, \hat{Q}^{i+k}(Q^{j-k}[1]),$$

$$(i_4)_*(Bt)_*(e_i \otimes e_j) = \sum ({i+j-k \atop i})\, Q^{i+j-k}[1] * Q^k[1],$$

since $Sq_*^k(e_j) = ({j-k \atop k})\, e_{j-k}$ and $(B\mu)_*(e_i \otimes e_{j-k}) = ({i+j-k \atop i})\, e_{i+j-k}$.

The summation formula above implies the formula

$$(i_4)_*(Bs)_*\left(\sum_{\ell=0}^{j} ({j+\ell \atop \ell})\, e_{i+\ell} \otimes e_{j-\ell}\right) = \hat{Q}^i(Q^j[1]).$$

Now using $(Bs)_* = (Bt)_*$ and once more 6.20 the result follows.

REMARK 6.22. The formulae above giving the homology interplay between the two structures on $Q(S^0)$ were originally developed in [73],

[139] and in papers by May and Tsuchiya (see [85] and [86] for complete references). The results were later conceptualized in the notion of E_∞ ring spaces by May, Quinn and Ray [85], and more recently in the notion of hyper Γ-spaces by R. Woolfson [144].

C. *The Pontrjagin ring* $H_*(BSG)$

In Chapter 3.F we calculated the Pontrjagin ring of $Q(S^0)$ under loop product. We now reinterpret the result (cf. 3.38) in terms of homology operations. Let $V'_n \subset \Sigma_{2^n}$ be the subgroup also used in the proof of 3.23 (and conjugate to the detecting subgroup V_n). Since

$$V'_n = Z/2\times V'_{n-1} \subset Z/2\times\Sigma_{2^{n-1}} \xrightarrow{1\times\Delta} Z/2\wr\Sigma_{2^{n-1}} \xrightarrow{J} \Sigma_{2^n}$$

(with $V'_1 = \Sigma_2$) the composition

$$BV'_n \longrightarrow B\Sigma_{2^n} \xrightarrow{i_{2^n}} Q(S^0)$$

maps $H_*(BV'_n)$ into the vector space spanned by all elements $Q_I[1] = Q_{i_1}\cdots Q_{i_n}[1]$. By 6.8 the elements Q_I with $i_1 \le \cdots \le i_n$ span the same vector space and by 6.6 (ii) $Q_I[1]$ is a loop square unless $i_1 > 0$. Now $BV'_n \to B\Sigma_{2^n}$ is homotopic to $BV_n \to B\Sigma_{2^n}$, and comparing with 3.38 a count of dimensions gives

COROLLARY 6.23.

$$H_*(Q(S^0)) = P\{Q_{i_1}\cdots Q_{i_n}[1] \mid 0 < i_1 \le \cdots < i_n\} \otimes Z/2[Z]\,.$$

The precise connection between the classes $E_{(j_1,\cdots,j_n)}$ from 3.38 and the classes $Q_I[1]$ above is somewhat complicated. However, a simple calculation shows that

$$E_{(j_1,\dots,j_n)} = Q_{i_1,\dots,i_n}[1] + \text{"longer" terms}$$

with $i_\nu = j_1 + \cdots + j_\nu$.

REMARK 6.24. The homology operations $Q^i (i \geq 0)$ modulo the Adem relations 6.8 and the excess relations $Q^i Q^j = 0$ for $i < j$ generate an associative algebra $\mathcal{R}$ (under composition). $\mathcal{R}$ is a Hopf algebra and the mod 2 homology of an infinite loop space is an algebra over $\mathcal{R}$. We have seen above that each element in $H_*(BV_n)$ defines a unique element of $\mathcal{R}$. Dually, each element of the dual algebra $\mathcal{R}^*$ defines an element of $H^*(BV_n)$. The structure of $\mathcal{R}^*$ was calculated in [73], adopting a procedure of Milnor [156]. The generators $\xi_{i,n} \in \mathcal{R}^*$ from [73] correspond to the classes D_{n-i}/D_n in 3.24. Thus we see that the connection between the classes $Q^I[1]$ and E_J is analogous to the connection between the Steenrod generators and the Milnor generators. For the mod 2 Steenrod algebra cf. [130], [156].

The homology ring $(H_*(SG), \circ)$ was calculated in Chapter 3.F. Using the generators $Q_I[1]$ the result takes the following form

$$\begin{aligned} H_*(SG) = {} & E\{Q_i[1] * [-1] \mid i \geq 1\} \otimes P\{Q_0 Q_i[1] * [-3] \mid i \geq 1\} \otimes \\ & P\{Q_I[1] * [1-2^n] \mid 0 < i_1 \leq \cdots \leq i_n,\ n \geq 2\}. \end{aligned}$$

For our purposes, this set of generators is unsuitable: we need the following result from [73],

THEOREM 6.25.

$$\begin{aligned} H_*(SG) = {} & E\{Q_i[1] * [-1] \mid i \geq 1\} \otimes P\{Q_0 Q_i[1] * [-3] \mid i \geq 1\} \otimes \\ & P\{\hat{Q}_{i_1} \cdots \hat{Q}_{i_{n-2}} (Q_{i_{n-1}} Q_{i_n}[1] * [-3]) \mid 0 < i_1 \leq \cdots \leq i_n,\ n \geq 2\}. \end{aligned}$$

Proof (Sketch). A detailed proof can be found in [73, pp. 258-265] or in [86, pp. 131-141], so we just give a rough outline of the argument. The problem is to calculate $\hat{Q}_i(Q_I[1] * [1-2^\ell])$ where $\ell \geq 2$ is the length of I. The mixed Cartan formula 6.15 gives

$$\hat{Q}_i(Q_I[1] * [1-2^\ell]) = Q_i Q_I[1] * [1-2^{\ell+1}] + \hat{Q}_i(Q_I[1]) * [1-2^{2\ell}] + \text{extra terms}.$$

One then shows that the extra terms (which are clearly decomposable in the loop product) are decomposable also in the composition product, and one is left with the term $\hat{Q}_i(Q_I[1]) * [1-2^{2\ell}]$.

In view of 6.18 each $Q_I[1]$ can be decomposed as

$$Q_I[1] = \sum Q_{j_1}[1] \circ \cdots \circ Q_{j_n}[1].$$

But $\hat{Q}_i$ satisfies a Cartan formula (cf. 6.6 (vi)) with respect to the composition product, so $\hat{Q}_i(Q_I[1])$ can be calculated from 6.21:

$$\hat{Q}_i(Q_I[1]) * [1-2^{2\ell}] = \sum Q_K[1] * [1-2^{2\ell}] + \text{extra terms}$$

where the summation runs over certain sequences K of the same length ℓ as I. Again the extra terms are decomposable in the $*$ product, and one must prove that they are also decomposable in the $\circ$ product. Then 6.25 follows immediately.

We end the chapter with a calculation of the Pontrjagin ring $H_*(BSG)$. A detailed argument can be found in [73, p. 266] or in [86, pp. 119-121]. The result is

THEOREM 6.26.

$$H_*(BSG) = H_*(BSO) \otimes$$
$$E\{\sigma_*(Q_0Q_i[1]*[-3]) \mid i \geq 1\} \otimes P\{\sigma_*(Q_iQ_j[1]*[-3]) \mid 0<i\leq j\} \otimes$$
$$P\{\hat{Q}_{i_1-1} \cdots \hat{Q}_{i_{n-2}-1}\sigma_*(Q_{i_{n-1}}Q_{i_n}[1]*[-3]) \mid 1 < i_1 \leq i_2 \leq \cdots \leq i_n,\ n>2\}$$

where $\sigma_*: H_*(SG) \to H_*(BSG)$ *is the suspension map induced from* $\Sigma\Omega BSG \to BSG$.

Proof (Sketch). We use the Eilenberg-Moore spectral sequence

$$\mathrm{Tor}_{H_*(SG)}(Z/2, Z/2) \Longrightarrow H_*(BSG),$$

but the reader unfamiliar with this might instead prefer to use the Serre spectral sequence. The spectral sequence above is dual to the spectral sequence used in 3.45, in particular it collapses, so $H_*(BSG)$ is additively equal to $\mathrm{Tor}_{H_*(SG)}(\mathbf{Z}/2, \mathbf{Z}/2)$. Thus from 6.25 we get (additively)

$$H_*(BSG) = H_*(BSO) \otimes E\{\sigma_*(Q_iQ_j[1]*[-3]) \mid 0 \leq i \leq j\} \otimes$$
$$E\{\sigma_*(\hat{Q}_{i_1} \cdots \hat{Q}_{i_{n-2}}(Q_{i_{n-1}} Q_{i_n}[1]*[-3])) \mid 0 < i_1 \leq \cdots \leq i_n,\ n>2\}.$$

Since $H_*(BSG)$ is a commutative and cocommutative Hopf algebra it is a truncated polynomial algebra by a well-known structure theorem ([101]). We check for truncations. First, by 6.6 (ii) and (iv)

$$\sigma_*(Q_0Q_i[1]*[-3])^2 = \sigma_*\hat{Q}_1(Q_0Q_i[1]*[-3]),$$

and $\hat{Q}_1(Q_0Q_i[1]*[-3]) = 0$; this follows by a routine calculation from 6.8, 6.15, 6.19 and 6.21. Hence $\sigma_*(Q_0Q_i[1]*[-3])$ is an exterior generator as claimed. Second, if $i>0$ then each $\hat{Q}_1 \cdots \hat{Q}_1(Q_iQ_j[1]*[-3])$ is a polynomial generator of $H_*(SG)$, so the elements $\sigma_*(Q_iQ_j[1]*[-3])$ generate a polynomial subalgebra when $i>0$. Finally, if $n>2$ then

$$\sigma_*(\hat{Q}_{i_1} \cdots \hat{Q}_{i_{n-2}}(Q_{i_{n-1}} Q_{i_n}[1]*[-3])) = \hat{Q}_{i_1-1} \cdots \hat{Q}_{i_{n-2}-1}(\sigma_*(Q_{i_{n-1}} Q_{i_n}[1]*[-3]))$$

and this is a square precisely when $i_1 = 1$. The elements with $i_1 > 1$ form polynomial generators. This completes the argument.

CHAPTER 7
THE 2-LOCAL STRUCTURE OF B(G/TOP)

In Chapter 4A we have seen that 2-locally G/TOP is a product of Eilenberg-MacLane spaces

$$G/TOP[2] \simeq \prod_{i=1}^{\infty} K(\mathbf{Z}_{(2)}, 4i) \times K(\mathbf{Z}/2, 4i-2) .$$

In this section we prove a similar theorem for its first two deloopings (in the Boardman-Vogt infinite loop space structure).

THEOREM 7.1. (i) $B(G/TOP)[2] \simeq \prod_{i=1}^{\infty} K(\mathbf{Z}_{(2)}, 4i+1) \times K(\mathbf{Z}/2, 4i-1)$.

(ii) $B^2(G/TOP)[2] = \prod_{i=1}^{\infty} K(\mathbf{Z}_{(2)}, 4i+2) \times K(\mathbf{Z}/2, 4i)$.

We also prove that

$$B(G/PL)[2] = E_3 \times \prod_{i=2}^{\infty} K(\mathbf{Z}_{(2)}, 4i+1) \times K(\mathbf{Z}/2, 4i-1)$$

where E_3 is the two-stage Postnikov system with non-trivial k-invariant $\beta(Sq^2(\iota))$ in $H^6(K(\mathbf{Z}/2, 3), \mathbf{Z})$, with a similar result holding for $B^2(G/PL)[2]$.

These results were first proved by the authors in [78]. They imply that the 2-local obstruction to the existence of a topological structure on a spherical fiber space is a set of ordinary cohomology classes. This was also obtained by

Brumfiel and Morgan [36] and Jones [152], and was originally proved under the assumption that the base space is 4-connected by Levitt and Morgan. The methods of these papers are quite different; they all use certain refinements of the "transversality obstruction" of Levitt.

In contrast to 7.1, the third delooping $B^3(G/TOP)[2]$ does not split as a product of Eilenberg-Maclane spaces. This relies on calculations given in [75], [76] (cf. 7.4 below).

The core of 7.1 is the splitting of $B(G/TOP)[2]$, and we proceed to outline the proof, as it is quite involved. We begin by showing that the $\mathbf{Z}_{(2)}$ integral k-invariants in degrees $4i+2$ must be zero if the lower k-invariants are all zero, and similarly for the $\mathbf{Z}/2$ k-invariants in degrees $4i$. The first step is obtained by using the fact that since the space G/TOP is an infinite loop space all the k-invariants are stable, and showing that the only possible stable k-invariant in dimension $4i+2$ has mod 2 reduction $\Sigma_I(Sq^I \iota)^2$. Then we observe that any time such a k-invariant occurs the Hopf algebra structure in $H^*(G/TOP; \mathbf{Z}/2)$ would fail to be primitively generated. This gives a contradiction.

The argument for the $\mathbf{Z}/2$-generators is more delicate, however. We begin by studying the Eilenberg-Moore spectral sequence converging to $H_*(B(G/TOP); \mathbf{Z}/2)$ and the relation of differentials to the Dyer-Lashof operations and Massey products. These results together with the results of [76] giving the action of the Dyer-Lashof operations in $H_*(G/TOP; \mathbf{Z}/2)$ then show that no differentials are possible in the Eilenberg-Moore spectral sequence so $E^2 = E^\infty$. This in turn implies the triviality of the $4i$ dimensional k-invariant.

A. *Products of Eilenberg-MacLane spaces and operations in* $H_*(G/TOP)$

We start out by recalling the structure of $H^*(K(\mathbf{Z}/2,n); \mathbf{Z}/2)$ and $H^*(K(\mathbf{Z}_{(2)},n); \mathbf{Z}/2)$. A sequence

$$Sq^I = Sq^{i_1} Sq^{i_2} \cdots Sq^{i_r}$$

in the Steenrod algebra is called *admissible* if $i_j \geq 2i_{j+1}$ for all j and it has excess

$$E(I) = i_1 - i_2 - \cdots - i_r .$$

These monomials form an additive basis for the Steenrod algebra and for the spaces in question we have ([123])

THEOREM 7.2. (i) $H^*(K(Z/2,m); Z/2) = P\{\cdots, Sq^I(\iota_m), \cdots\}$ *where* I *runs over all admissible sequences with* $E(I) < m$.

(ii) $H^*(K(Z_{(2)},m); Z/2) = P\{\cdots, Sq^I(\iota_m), \cdots\}$, I *admissible,* $E(I) < m$ *and* $i_r > 1$.

For any abelian group Γ, $K(\Gamma,m)$ has unique deloopings: if $\Omega^r(X) = K(\Gamma,m)$ then $X \simeq K(\Gamma,m+r)$ as one sees by checking the homotopy groups.

COROLLARY 7.3. *In* $H_*(K(Z/2,m); Z/2)$ *and* $H_*(K(Z_{(2)},m); Z/2)$ *the Dyer-Lashof operations* Q_i *are identically zero for* $i \geq 0$.

Proof. Let $\mu: K(Z/2,m) \times K(Z/2,m) \to K(Z/2,m)$ be the H-structure. Then we must have

$$\mu^*(\iota_m) = \iota_m \otimes 1 + 1 \otimes \iota_m$$

and therefore, by naturality

$$\mu^*(Sq^I(\iota_m)) = Sq^I(\iota_m) \otimes 1 + 1 \otimes Sq^I(\iota_m)$$

showing that $H^*(K(Z/2,m); Z/2)$ is primitively generated. By a general result on Hopf algebras ([101]) the vector space of primitive elements of $H^*(K(Z/2,m); Z/2)$ is generated by the powers $(Sq^I(\iota_m))^{2^r}$. In other words the $Sq^I(\iota_m)$ with I admissible and $E(I) \leq m$ form a basis for $PH^*(K(Z/2,m); Z/2)$. Let

$$\sigma^*: H^*(K(Z/2,m+1); Z/2) \to H^*(K(Z/2,m); Z/2)$$

be the suspension, induced from $\Sigma\Omega K(Z/2,m+1) \to K(Z/2,m+1)$. We have seen that σ^* restricts to a surjection on the primitive elements. Dualizing to homology, primitives are replaced by indecomposables so

$$\sigma_*: QH_*(K(Z/2,m); Z/2) \to QH_*(K(Z/2,m+1); Z/2)$$

is injective. The result now follows from 6.6. Indeed if Q_i is not identically zero, then there must be a primitive indecomposable α with $Q_i(\alpha) \neq 0$. If $Q_i(\alpha)$ is indecomposable, then iterating σ_* i times we have

$$\sigma_*^{(i)} Q_i(\alpha) = Q_0(\sigma_*^{(i)}(\alpha)) = (\sigma_*^{(i)}(\alpha))^2$$

which contradicts the injectivity of σ_* above. If $Q_i(\alpha)$ is decomposable, then as it is primitive it must be a square. But $H_*(K(Z/2,m);Z/2)$ is an exterior algebra, and 7.3 follows.

REMARK. In Chapter 1 we described models for $K(\Gamma,m)$ which are abelian monoids. Since

$$d_2 : E\Sigma_2 \times_{\Sigma_2} K(\Gamma,m)^2 \to K(\Gamma,m)$$

measures the deviation from strict commutativity it is clear (at least heuristically) that $Q_i \equiv 0$ for $i>0$. Indeed, it is not hard to give a noncalculational proof of 7.3 along these lines.

For a product of Eilenberg-MacLane spaces, the situation is more complicated since deloopings are no longer unique. Consider, for example, the two stage Postnikov system E,

$$E \longrightarrow K(Z/2,2) \xrightarrow{\beta\, Sq^2} K(Z_{(2)},5)\ .$$

Then $\Omega E = K(Z/2,1) \times K(Z_{(2)},3)$ but $E \not\simeq K(Z/2,2) \times K(Z_{(2)},4)$. This situation occurs for the space G/TOP: even though it is a product of Eilenberg-MacLane at the prime 2 it does not have the same loop space structure as we see from

THEOREM 7.4. *In* $H_*(G/TOP; Z/2)$, $Q_0 \equiv 0$ *and* $Q_1 \equiv 0$, *but*

$$\langle Q_2(x_{4i+2}), K_{8i+6}\rangle = \langle x_{4i+2}, K_{4i+2}\rangle$$

for any $x_{4i+2} \in H_{4i+2}(G/TOP; Z/2)$. *Here* K_{4i+2} *denotes the "Kervaire" fundamental classes in* $H^{4i+2}(G/TOP; Z/2)$ (*cf. Theorem 4.9*).

This theorem is proved in [75], [76]. The idea of the proof is to use the surgery interpretation of maps into G/TOP. Suppose $f: M^{4i+2} \to G/TOP$ is a given singular manifold of G/TOP, then the steps leading to 7.4 are

(1) If $\phi: \tilde{M}^{4i+2} \to M^{4i+2}$ is the degree 1 normal map corresponding to f then

$$1 \times \phi \times \phi: S^n \times_{Z/2} \tilde{M} \times \tilde{M} \to S^n \times_{Z/2} M \times M$$

is the degree 1 normal map associated to

$$\Gamma_{n+1}(M) \longrightarrow \Gamma_{n+1}(G/TOP) \xrightarrow{d_2} G/TOP$$

where d_2 is the structure map associated with the infinite loop space structure on G/TOP defined by Boardman-Vogt [19], and Γ_{n+1} is the quadratic construction.

(2) The Kervaire surgery obstruction for ϕ is the same as the obstruction for

$$1 \times \phi \times \phi: S^2 \times_{Z/2} \tilde{M} \times \tilde{M} \to S^2 \times_{Z/2} M \times M .$$

(3) Apply the Sullivan formula,

$$s_K(M, \phi) = \langle f^*(\Sigma K_{4j-2}) \cup v^2, [M] \rangle .$$

Only (2) provides any real difficulties and its proof depends on the homotopy theoretic methods for evaluating the Kervaire invariant (cf. [27]).

We proceed to examine the subgroup of primitive elements, $PH^*(X; Z_{(2)})$ when X is an H-space whose underlying homotopy type is a product of $K(Z/2,j)$'s and $K(Z_{(2)},j)$'s. Our main tool is the mod 2 Bockstein spectral sequence (see [23]),

$$E_1(X) = H^*(X; Z/2)$$

$$E_\infty(X) = H^*(X; Z_{(2)})/\text{Tor} .$$

When X is an H-space then $(E_r(X), d_r)$ is a spectral sequence of Hopf algebras. Let $j_r : H^*(X; Z/2^r) \to E_r(X)$ denote the reduction homomorphism. It is a surjection with kernel $2^*H^*(X; Z/2^{r-1}) + \rho_r\beta_{r-1}H^*(X; Z/2^{r-1})$, where 2^* is induced from the inclusion $Z/2^{r-1} \subset Z/2^r$, β_{r-1} is the integral Bockstein homomorphism associated with the coefficient sequence $0 \to Z_{(2)} \xrightarrow{2^{r-1}} Z_{(2)} \to Z/2^{r-1} \to 0$ and ρ_r is the reduction to $Z/2^r$ coefficients. If $j_r(x) \neq 0$ then x has order 2^r in $H^*(X; Z/2^r)$. We observe

LEMMA 7.5. *Let* $x \in H^*(X; Z/2)$, *where* X *is any* H-*space. Then* $2^{r-1}x$ *is primitive if and only if* $j_r(x)$ *is primitive.*

We review the Bockstein spectral sequence of a single $K(\Lambda,n)$ where $\Lambda = Z_{(2)}$ or $Z/2$ (cf. [23]). Let $E_r\{x\}$ denote the "model" spectral sequence with

$$E_{r+2}\{x\} = P\{x^{2^r}\} \otimes E\{yx^{2^r-1}\}, \ \deg(x) = 4n, \ \deg(y) = 4n+1$$

$$d_{r+2}(x^{2^r}) = yx^{2^r-1}$$

and let $E_r\{x_1, x_2, \cdots\}$ be the tensor product of the individual $E_r\{x_i\}$. Then for $r \geq 2$,

$$\text{(i)} \quad E_r(K(Z/2,n)) = E_r\{\cdots, x_i, \cdots\}$$

7.6 $$\text{(ii)} \quad E_r(K(Z_{(2)},2n)) = P\{\iota_{2n}\} \otimes E_r\{\cdots, x_i, \cdots\}$$

$$\text{(iii)} \quad E_r(K(Z_{(2)},2n-1)) = E\{\iota_{2n-1}\} \otimes E_r\{\cdots, x_i, \cdots\}$$

where ι_{2n} and ι_{2n-1} are reductions of integral primitive elements. The number of factors in each of the cases above as well as the naming of the elements x_i in $E_1(K(\Lambda,n)) = H^*(K(\Lambda,n); Z/2)$ is available from 7.2 but not relevant to our purpose. We do, however, need the fact that each $x_i \in H^{4*}(K(\Lambda,n); Z/2)$ is the square of a primitive (indecomposable) element.

LEMMA 7.7. *The subgroup of primitive torsion elements in* $H^*(K(\Lambda,n);Z_{(2)})$ *forms a vector space over* $Z/2$. *More precisely,* $\operatorname{Tor} PH^*(K(\Lambda,n);Z_{(2)})$ *is spanned by elements of one of two types:*

(i) $2^{r-1}\beta_r(z_r)$ *where* $z_r \in H^*(K(\Lambda,n);Z/2^r)$ *has reduction* $j_r(z_r) = x_i^{2^r}$ *in* $E_r(K(\Lambda,n))$.

(ii) $\beta_1(y)^{2^a}$ *for* $y \in PH^*(K(\Lambda,n);Z/2)$.

Proof. It is a consequence of 7.5 that the elements $2^{r-1}\beta_r(z_r)$ are primitive. It suffices to prove that a primitive torsion element p is a linear combination of the elements listed in (i) and (ii). Suppose inductively that

$$q_r = p - \sum_{i=1}^{r-1} 2^{i-1}\beta_i(z_i)$$

is divisible by 2^{r-1} in $H^*(K(\Lambda,n);Z_{(2)})$. From 7.5 we have that $j_r((1/2^{r-1})q_r)$ is primitive and from 7.6 that there is an element z_r with $j_r((1/2^{r-1})q_r) = j_r(\beta_r(z_r))$. But then $q_r - 2^{r-1}\beta_r(z_r)$ reduces to zero in $H^*(K(\Lambda,n);Z/2^r)$ and is therefore divisible by 2^r. This process ends after a finite number of steps since $K(\Lambda,n)$ is of finite type. We finally note that if $j_1(p) = (\beta_1(y))^{2^a}$ for $a > 0$ then $p = \beta_1(y)^{2^a}$ since the elements $\beta_r(z_r)$ for $r > 1$ all have dimension congruent to 1 (mod 4). This completes the proof.

A product of Eilenberg-MacLane spaces can have several H-space structures. Let $E_{4,k}$ be the fiber in the fibration

$$E_{4,k} \longrightarrow K(Z/2, k+3) \xrightarrow{Sq^4} K(Z/2, k+7)\ .$$

Then $\Omega E_{4,k} = E_{4,k-1}$. In particular, $E_{4,0}$ has the homotopy type of $K(Z/2,3) \times K(Z/2,6)$. The H-space structure on $E_{4,0}$ however, is distinct from the ordinary structure on the product, since in $H^*(E_{4,0}; Z/2)$,

$$\psi(\iota_6) = \iota_6 \otimes 1 + \iota_3 \otimes \iota_3 + 1 \otimes \iota_6 .$$

(Compare [1], [93].)

More generally, if X is an H-space which is homotopy equivalent to a product of $K(Z/2,i)$'s and $K(Z_{(2)},j)$'s and if

$$K(Z/2, 4n+1) \xrightarrow{j} E \xrightarrow{i} X \xrightarrow{\pi} K(Z/2, 4n+2)$$

is a fibration sequence with $\pi^*(\iota_{4n+2}) = Sq^{2n+1}(x)$ for some primitive element $x \in H^{2n+1}(X; Z/2)$, then in $H^*(\Omega E; Z/2)$ there is a class ι_{4n} with $j^*(\iota_{4n})$ the generator of $H^{4n}(K(Z/2, 4n); Z/2)$ and such that

7.8 $$\bar{\psi}(\iota_{4n}) = \sigma^*(i^*(x)) \otimes \sigma^*(i^*(x)) ,$$

where $\bar{\psi}$ is the reduced diagonal. This follows easily using the methods of [93] or [153].

Let A be an algebra over $Z/2$ equipped with two coalgebra structures ψ_1 and ψ_2 and such that the (A, ψ_i) are commutative and cocommutative Hopf algebras. Further, suppose that (A, ψ_2) is primitively generated. (A, ψ_1) is a tensor product of monogenic Hopf algebras by a theorem of Milnor and Moore [101]. Moreover, the primitive elements of (A, ψ_1) are contained among the indecomposables and elements of the form x^{2^i} with x primitive. We conclude that the primitive elements of (A, ψ_1) occur in a subset of the same dimensions as the primitive elements of (A, ψ_2). As a corollary of the proof of 7.7 we now have

LEMMA 7.9. *Let* X *be a homotopy commutative* H-*space and suppose the underlying space has the homotopy type of a product of Eilenberg-MacLane spaces,* $X \simeq \Pi K(\Lambda, j)$, $\Lambda = Z_{(2)}$ *or* $Z/2$. *Then a primitive torsion element of* $H^*(X; Z_{(2)})$ *either occurs in dimension* $4t+1$ *or it has a non-zero* $Z/2$ *reduction.*

B. *Massey products in infinite loop spaces*

We now introduce Massey products in the homology of infinite loop spaces and relate them to the kernel of the suspension map $\sigma_*: H_*(\Omega X) \to H_*(X)$.

Suppose we have a chain complex C_* together with an associative product on C_* with respect to which ∂ is a derivation, $\partial(a \cdot b) = \partial a \cdot b + (-1)^{|a|} a \cdot \partial b$. For example, C_* could be the singular chain complex of an associative H-space or the singular cochain complex of any space Y (with the usual diagonal approximation).

Under these circumstances $H(C_*, \partial)$ is also an algebra and Massey products arise when there are more relations in $H(C_*, \partial)$ than the ring structure of C_* would directly account for.

The simplest case is the 3-fold Massey product $\langle a, b, c \rangle$ which is defined for any triple (a, b, c) of elements in $H_*(C_*, \partial)$ satisfying $ab = bc = 0$. Its definition is

DEFINITION 7.10. Let (a, b, c) satisfy $ab = bc = 0$ in $H_*(C_*, \partial)$, then $\langle a, b, c \rangle$ consists of all classes $x \in H(C_*, \partial)$ which can be represented by chains of the form

$$A\{c\} - (-1)^{|a|}\{a\}B$$

where $\{a\}$, $\{b\}$ and $\{c\}$ are in C_* and represent a, b, c respectively while $\partial A = \{a\} \cdot \{b\}$, $\partial B = \{b\} \cdot \{c\}$.

If a, b, c have dimensions i, j, k respectively and ∂ has degree ε, then

$$\langle a, b, c \rangle \subset H_{i+j+k-\varepsilon}(C_*, \partial)$$

and the following lemma is a formal manipulation in C_*.

LEMMA 7.11. *Let* $x \in \langle a, b, c \rangle$ *then*

$$x + [a \cdot H_{j+k-\varepsilon}(C_*, \partial) + H_{i+j-\varepsilon}(C_*, \partial) \cdot c] = \langle a, b, c \rangle .$$

The elements in the brackets are said to represent the indeterminacy of $\langle a, b, c \rangle$.

REMARK 7.12. In cohomology the 3-fold Massey product has a curious connection with geometry in that it gives a method to show that 3 disjointly embedded circles S_1, S_2, S_3 in S^3 are linked even if any two are unlinked. For example, consider the 'Ballantine symbol'

7.13

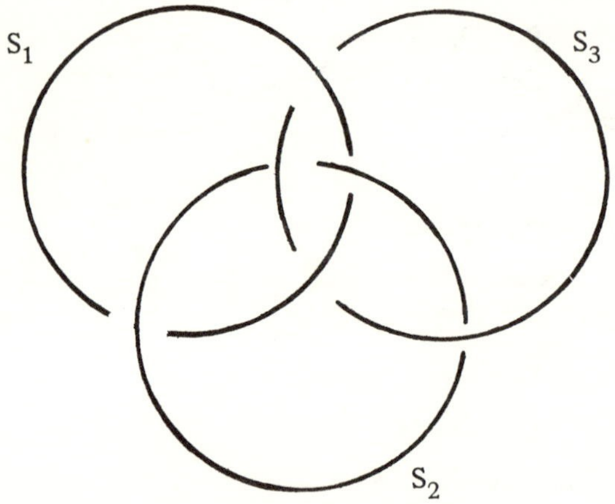

By Alexander duality $H^1(S^3-(S_1 \cup S_2 \cup S_3), \mathbf{Z}) = \mathbf{Z}^{(3)}$ with generators a, b, c Poincaré dual to S_1, S_2, S_3 respectively.

If S_1, S_2 are disjointly embedded in S^3 then $H^1(S^3-(S_1 \cup S_2), \mathbf{Z}) = \mathbf{Z}^{(2)}$ with generators α, β dual to S_1, S_2 respectively. Also $H^2(S^3-S_1 \cup S_2, \mathbf{Z}) = \mathbf{Z}$ with generator e and from geometry if n is the linking number of S_1, S_2, then $\alpha \cup \beta = \pm ne$.

Using the 3 inclusions

$$S^3 - S_1 \cup S_2 \cup S_3 \to S^3 - (S_{i_1} \cup S_{i_2})$$

we see that for 7.13 $ab = ac = bc = 0$ in $H^*(S^3-S_1 \cup S_2 \cup S_3, \mathbf{Z})$. Hence, $\langle a, b, c\rangle$ and $\langle a, c, b\rangle$ are both defined with zero indeterminacy in $H^2(S^3-S_1 \cup S_2 \cup S_3, \mathbf{Z}) = \mathbf{Z}^{(2)}$. Moreover in [87] it is shown that $\langle a, b, c\rangle$ is non-zero. This detects the non-triviality of 7.13. In fact $\langle a, b, c\rangle$ and $\langle a, c, b\rangle$ give a basis for $H^2(\ ;\mathbf{Q})$.

It should be remarked that we could actually form 6 Massey products using a, b, c, however, there are two identities connecting Massey products, a Jacobi relation

$$0 \in \langle a, b, c\rangle \pm \langle c, a, b\rangle \pm \langle b, c, a\rangle$$

and a symmetry relation

$$0 \in \langle a, b, c\rangle \pm \langle c, b, a\rangle$$

which imply that there are at most 2 independent products obtained this way.

Now we define the higher Massey products and the somewhat more general matric Massey products.

DEFINITION 7.14. Let (C_*, ∂) be as before. Let M and N be $n\times m$ (respectively $m\times k$) matrices with entries in $H(C_*, \partial)$. We say that M and N are multipliable if $\deg(m_{ij}) + \deg(n_{jk})$ depends only on i and k.

When M and N are multipliable matrices, then $M\cdot N$ is an $n\times k$ matrix with entries in $H(C_*, \partial)$.

DEFINITION 7.15. Let $M_1, \cdots, M_n$ be a set of matrices in $H(C_*, \partial)$ with M_1 of type $1\times s$ and M_n of type $t\times 1$, and such that M_i, M_{i+1} are multipliable for all i. The n-fold matric Massey product $\langle M_1, \cdots, M_n\rangle$ is said to be defined if there exist matrices $N_{i,j}$, $1 \le i \le j \le n+1$, $1 \le j-i \le n-1$ with entries in C_* which satisfy

$$\partial N_{i,i+1} = 0, \qquad \partial N_{i,j} = \sum_k N_{i,k} N_{k,j}$$

and $\mathrm{cls}\{N_{i,i+1}\} = M_i$ in homology. The matric Massey product is then the set of all classes in $H(C_*, \partial)$ represented by cycles of the form $\sum_k N_{1,k} N_{k,n+1}$.

REMARK 7.16. A naturally arising example of a multipliable system is obtained by considering $Sq^i(a \cup b) = \sum Sq^r(a) \cup Sq^{i-r}(b)$ in ordinary cohomology when we rewrite the formula above as

7.17 $$Sq^i(a\cup b) = (Sq^i(a), Sq^{i-1}(a), \cdots, Sq^0(a)) \begin{pmatrix} Sq^0(b) \\ Sq^1(b) \\ \cdot \\ \cdot \\ Sq^i(b) \end{pmatrix}.$$

Again, if we expand out $Sq^i(a \cup b \cup c)$ by the Cartan formula we obtain

$$7.18 \quad Sq^i(a\cup b\cup c) = (Sq^i(a), \cdots, Sq^0(a)) \begin{pmatrix} Sq^0(b) & 0 & \cdots & 0 \\ Sq^1(b) & Sq^0(b) & \cdots & 0 \\ \vdots & \vdots & & \vdots \\ Sq^i(b) & Sq^{i-1}(b) & \cdots & Sq^0(b) \end{pmatrix} \begin{pmatrix} Sq^0(c) \\ \cdot \\ \cdot \\ \cdot \\ Sq^i(c) \end{pmatrix}.$$

In the homology of an infinite loop space the Dyer-Lashof operations satisfy 6.6 (v) so formulae analogous to 7.17, 7.18 are valid with Sq^i replaced by Q^i. In this context the following theorem is a special case of the main theorem, Theorem 0 of [89], upon using the results of [72] to provide the homotopies required in the proof of Theorem 0.

THEOREM 7.19. *Let* X *be an infinite loop space and suppose* $x \in H_*(X; Z/2)$ *is an element of the matric Massey product* $\langle M_1, \cdots, M_n \rangle$. *Then* $Q_2(x)$ *is contained in the* n*-fold Massey product*

$$\left\langle (Q_2(M_1), Q_1(M_1), Q_0(M_1)), \begin{pmatrix} Q_0(M_2) & 0 & 0 \\ Q_1(M_2) & Q_0(M_2) & 0 \\ Q_2(M_2) & Q_1(M_2) & Q_0(M_2) \end{pmatrix}, \cdots, \begin{pmatrix} Q_0(M_n) \\ Q_1(M_n) \\ Q_2(M_n) \end{pmatrix} \right\rangle$$

where $[Q_s(M_r)]_{i,j} = Q_s((M_r)_{i,j})$.

REMARK 7.20. It is evident from the definition that a matric Massey product

$$\left\langle (Q_2(M_2), 0, 0), \begin{pmatrix} 0, & 0, 0 \\ 0, & 0, 0 \\ Q_2(M_2), & 0, 0 \end{pmatrix}, \cdots, \begin{pmatrix} 0, & 0, 0 \\ 0, & 0, 0 \\ Q_2(M_{n-1}), & 0, 0 \end{pmatrix}, \begin{pmatrix} 0 \\ 0 \\ Q_2(M_n) \end{pmatrix} \right\rangle$$

is always defined and always contains 0. In particular, from 7.4 it follows that if $x \in H_*(G/TOP; Z/2)$ belongs to an n-fold matric Massey product then $Q_2(x)$ belongs to an n-fold matric Massey product which also contains 0. Thus, since the indeterminacy of an n-fold Massey product consists of (n–1)-fold Massey products (see e.g. [89], pp. 41 and 42 for examples of these indeterminacy products) it follows that $Q_2(x)$ is contained in a strictly shorter Massey product.

COROLLARY 7.21. *Let* K_{4i+2} *be the Kervaire class in* $H^{4i+2}(G/TOP, Z/2)$, *then if* x *is contained in any matric Massey product it follows that* $\langle x, K_{4i+2}\rangle = 0$.

Proof. The class K_{4i+2} constructed in the proof of 4.9 can be shown to be primitive as is done, for example in [117]. Thus, it vanishes on homology decomposables. Now suppose x is an n-fold Massey product with $\langle x, K_{4i+2}\rangle = 1$. Then after at most n–2 iterations we have $1 = \langle Q_2 \cdots Q_2 x, K_{2^{j+2}(i+1)-2}\rangle$ from 7.4. But by 7.20, $Q_2 \cdots Q_2(x)$ can be assumed to be a 2-fold Massey product and hence a sum of decomposables and this is a contradiction.

Matric Massey products are very closely tied to the suspension map

$$\sigma_* : H_*(\Omega X; Z/2) \to H_*(X; Z/2)$$

by theorems of Kraines and May [67], [83]. The result we need is

THEOREM 7.22 (May). *The kernel of* σ_* *is exactly the set of all matric Massey products in* $H_*(\Omega X; Z/2)$.

COROLLARY 7.23. $K_{4i+2} \in \mathrm{Image}(\sigma^*)$ *for all* i *under the suspension map* $\sigma^* : H^{4i+3}(B(G/TOP); Z/2) \to H^{4i+2}(G/TOP; Z/2)$.

REMARK 7.24. The reader familiar with the Eilenberg-Moore spectral sequence might check [72] for an alternative proof of 7.23. It uses the fact that Dyer-Lashof operations act in the Eilenberg-Moore spectral sequence

$$\mathrm{Tor}_{H_*(\Omega X; Z/2)}(Z/2, Z/2) \Longrightarrow H_*(X; Z/2)$$

where X is an infinite loop space and 7.4 to show that for $X = B(G/TOP)$ the spectral sequence collapses.

C. *The proof of Theorem 7.1*

Consider a Postnikov decomposition of $B(G/TOP)[2]$,

$$\begin{array}{ccccccccc} \cdots \longrightarrow BE_5 \longrightarrow & BE_4 & \longrightarrow & BE_3 & \longrightarrow & BE_2 & \longrightarrow & K(\mathbf{Z}/2,3) \\ & \downarrow{\scriptstyle \pi_4} & & \downarrow{\scriptstyle \pi_3} & & \downarrow{\scriptstyle \pi_2} & & \downarrow{\scriptstyle \pi_1} \\ & K(\mathbf{Z}/2,12) & & K(\mathbf{Z}_{(2)},10) & & K(\mathbf{Z}/2,8) & & K(\mathbf{Z}_{(2)},6). \end{array}$$

It is completely determined by the k-invariants $k_r = \pi_r^*(\iota_{2r+4})$ in $H^*(BE_r; \pi_*(G/TOP))$. Since G/TOP is an infinite loop space, the same is true of each stage BE_r. In particular k_r must be in the image of the suspension map and hence primitive.

The proof of 7.1 (i). The proof is by induction over the Postnikov decompositions above. Suppose that the r'th stage BE_r has the homotopy type of a product of Eilenberg-MacLane spaces. We must show that the k-invariant in the next stage is zero. The k-invariant is determined by the first dimension in which the projection $\pi: B(G/TOP)[2] \to BE_r$ is not a homotopy equivalence, and is non-zero only if

$$\pi^*: H^{s+1}(BE_r; \pi_s(B(G/TOP)[2])) \to H^{s+1}(B(G/TOP); \pi_s(B(G/TOP)[2]))$$

is not injective. In our case the kernel must be cyclic with a primitive generator.

If $s = 4i+1$, we require a primitive element k_r of $H^{4i+2}(BE_r; \mathbf{Z}_{(2)})$ and from 7.9 either $k_r = 0$ or $j_1(k_r) \neq 0$ in $H^{4i+2}(BE_r; \mathbf{Z}/2)$. In the latter case, consider $\sigma^*(j_1(k_r))$. It is surely zero since G/TOP is a product of Eilenberg-MacLane spaces. Hence $j_1(k_r) = y^2$ for some primitive element y. This follows from the exact sequence (see [101])

$$0 \longrightarrow PH^*(BE_r; \mathbf{Z}/2) \xrightarrow{\xi} PH^*(BE_r; \mathbf{Z}/2) \longrightarrow QH^*(BE_r; \mathbf{Z}/2)$$

where ξ is the squaring map, indeed the suspension map from $QH^*(BE_r; Z/2)$ to $PH^*(E_r; Z/2)$ is injective by the inductive assumption. But y is odd dimensional, hence indecomposable, and thus $\sigma^*(y) \neq 0$ in $H^*(E_r; Z/2)$. In this case we would have a class ι_{4i} in $H^*(G/TOP; Z/2)$ with

$$\bar{\psi}(\iota_{4i}) = \sigma^*(y) \otimes \sigma^*(y)$$

(cf. 7.8). This contradicts the fact that $H^*(G/TOP; Z/2)$ is primitively generated. (Cf. 4.34.)

If $s = 4i-1$, then the possible k-invariant k_r belongs to $H^{4i}(BE_r; Z/2)$. That this must be zero follows from 7.23. This completes the proof.

Before we can give the proof of 7.1 (ii) we need a few preliminary remarks on the Eilenberg-Moore spectral sequence

$$\mathrm{Ext}_{H_*(X;Z/2)}(Z/2, Z/2) \Longrightarrow H^*(BX; Z/2)$$

when X is an infinite loop space. First, recall that the spectral sequence is associated to a natural geometric filtration

$$B_1X \subset B_2X \subset \cdots \subset B_nX \subset \cdots$$

of BX ([115], compare also the footnote to Theorem 1.5). In particular, the spectral sequence admits an action of the mod 2 Steenrod algebra. Second, a result of A. Clark [150] asserts that the spectral sequence admits a Hopf algebra structure. Hence, the differential of a primitive element is again primitive.

The proof of Theorem 7.1 (ii). From 7.1 (i) and 7.2 $H^*(B(G/TOP); Z/2)$ is a polynomial algebra on primitive generators. Thus $H_*(B(G/TOP); Z/2)$ is an exterior algebra and

$$\mathrm{Ext}_{H_*(B(G/TOP);Z/2)}(Z/2, Z/2) = P\{[p] \mid p \in PH^*(B(G/TOP), Z/2)\}.$$

The elements [p] of the E_2-term are primitive and have total degree $\deg(p)+1$.

We claim that all differentials vanish. Indeed, using the action of the Steenrod operations in the spectral sequence it suffices to see that the elements $[\iota_{2s+1}]$ are infinite cycles where ι_{2s+1} denotes the fundamental class of $H^{2s+1}(B(G/TOP); \mathbf{Z}/2)$. But $d_r([\iota_{2s+1}])$ is a primitive (since $[\iota_{2s+1}]$ is), it has odd total degree and filtration degree greater than one, so it must be zero. The collapse of the spectral sequence implies that the suspension homomorphism

$$\sigma^* : QH^*(B^2(G/TOP); \mathbf{Z}/2) \to PH^*(B(G/TOP); \mathbf{Z}/2)$$

is an isomorphism.

For $B^2(G/TOP)[2]$ the possible k-invariants occur in dimensions $4s+1$ and $4s+3$. Let B^2E_r denote the r'th stage in the Postnikov decomposition for $B^2(G/TOP)[2]$, and assume it is a product of Eilenberg-MacLane spaces. Then the r'th k-invariant is a primitive element in either $H^{4s+3}(B^2E_r; \mathbf{Z}_{(2)})$ or in $H^{4s+1}(B^2E_r; \mathbf{Z}/2)$. In the first case k_r is non-zero only if its mod 2 reduction $j_1(k_r)$ is non-zero (cf. 7.9). But $j_1(k_r)$ is odd-dimensional and primitive, hence indecomposable, and $\sigma^*(j_r(k_r)) = 0$ then implies that $j_1(k_r) = 0$. In the second case a similar remark applies. This completes the proof.

As mentioned earlier there are similar results for $B(G/PL)$ and $B^2(G/PL)$. Indeed, let E_3 and BE_3 be the fibers in the fibrations

$$E_3 \longrightarrow K(\mathbf{Z}/2, 3) \xrightarrow{\beta_1 Sq^2} K(\mathbf{Z}_{(2)}, 6)$$

$$BE_3 \longrightarrow K(\mathbf{Z}/2, 4) \xrightarrow{\beta_1 Sq^2} K(\mathbf{Z}_{(2)}, 7) .$$

THEOREM 7.25. *There are homotopy equivalences*

$$B(G/PL)[2] \simeq E_3 \times \prod_{n=2}^{\infty} K(Z_{(2)}, 4n+1) \times K(Z/2, 4n-1)$$

$$B^2(G/PL)[2] \simeq BE_3 \times \prod_{n=2}^{\infty} K(Z_{(2)}, 4n+2) \times K(Z/2, 4n) .$$

Proof. Consider the fibration

$$K(Z/2,4) \to B(G/PL) \to B(G/TOP) .$$

It is a fibering in the category of infinite loop spaces and thus classified by a stable mapping

$$B(G/TOP) \xrightarrow{\lambda} K(Z/2,5) .$$

In particular B(G/PL) is the fiber of λ. But

$$PH^5(B(G/TOP); Z/2) = Z/2 \oplus Z/2$$

with generators $Sq^2(\iota_3)$ and $\rho_1(\iota_5)$, respectively. Moreover, in view of the known structure of G/PL (cf. 4.8) the only possibility for $\lambda^*(\iota)$ is $\lambda^*(\iota) = Sq^2(\iota_3) + \rho_1(\iota_5)$, and the result on B(G/PL) easily follows. The result for $B^2(G/PL)$ is shown in a similar fashion.

CHAPTER 8

THE TORSION FREE STRUCTURE OF THE ORIENTED COBORDISM RINGS

In this chapter we begin our analysis of the oriented PL and topological cobordism rings, Ω_*^{PL} and Ω_*^{TOP}. The torsion structure of these groups is exceedingly involved, and not completely known at odd primes. Torsion questions are taken up in Chapter 14. But even the torsion free structure $\Omega_*^{PL}/\mathrm{Tor}$ $(=\Omega_*^{TOP}/\mathrm{Tor})$ is complicated. For example, $\Omega_*^{PL}/\mathrm{Tor}$ is not a polynomial ring. Geometrically, one may for each prime p construct fairly explicit PL manifolds whose cobordism classes give a minimal set of generators for $\Omega_*^{PL}/\mathrm{Tor} \otimes \mathbf{Z}_{(p)}$, but the manifolds will vary with p. This makes it difficult to list a minimal set of generators for the integral ring, $\Omega_*^{PL}/\mathrm{Tor}$. In this chapter we give fairly explicit constructions of a sufficient set of manifolds to generate $\Omega_*^{PL}/\mathrm{Tor}$, but leave the questions of minimal generating set, the precise algebraic structure and relation to characteristic classes to later chapters.

A. *The map* $\eta: \Omega_*(G/PL) \to \Omega_*^{PL}$

Given a singular (smooth) manifold in G/PL, $f: M \to G/PL$, there is an associated degree 1 normal map

$$\pi: \tilde{M} \to M, \quad \hat{\pi}: \nu_{\tilde{M}} \to \nu_M - \xi$$

where ξ is the PL-bundle represented by the composition $M \xrightarrow{f} G/PL \xrightarrow{\hat{j}} BPL$ (cf. Chapter 2.C). Moreover, if (M, f) is bordant to (M′, g), then there is a normal bordism covering the bordism above

$$\begin{array}{ccccc} \tilde{M} & \subset & \tilde{W} & \supset & \tilde{M}' \\ \downarrow{\scriptstyle \pi} & & \downarrow & & \downarrow{\scriptstyle \pi'} \\ M & \subset & W & \supset & M' \end{array}$$

where $\partial W = M \cup M'$, $\partial \tilde{W} = \tilde{M} \cup \tilde{M}'$. In general $\tilde{M}$ is a PL-manifold which is not even bordant to a differentiable one, and we set

8.1 $$\eta\{M, f\} = \{\tilde{M}\} .$$

The class $\{\tilde{M}\}$ depends only on the class of $\{M, f\}$ in $\Omega_*(G/PL)$ and hence defines the desired homomorphism

8.2 $$\eta : \Omega_*(G/PL) \to \Omega_*^{PL}(pt) .^{*)}$$

This is an Ω_* module map since the surgery problem over the composite map $M \times N \longrightarrow M \xrightarrow{f} G/PL$ is $\tilde{M} \times N \xrightarrow{1\times\pi} M \times N$. More generally, since G/PL is an H-space there is a product $\Omega_*(G/PL) \otimes \Omega_*(G/PL) \xrightarrow{\mu} \Omega_*(G/PL)$ and $\eta\mu(a \otimes \beta) = \eta(a) \cdot \eta(\beta)$.

We can give a homotopy theoretic description of the homomorphism η as follows. Consider the composition

$$\theta : BSO \times G/PL \xrightarrow{1\times\chi} BSO \times G/PL \xrightarrow{\tilde{j}\times\hat{j}} BSPL \times BSPL \xrightarrow{\oplus} BSPL$$

where χ is the automorphism $x \to x^{-1}$ (which exists for any loop space, cf. Chapter 3.F) and $\tilde{j}, \hat{j}$ are the natural maps. If $\gamma_{SPL}, \gamma_{SO}$ are the universal bundles then

$$\theta^*(\gamma_{SPL}) = \gamma_{SO} \times (-\gamma_{G/PL}), \quad \gamma_{G/PL} = \hat{j}^*(\gamma_{SPL}) .$$

But $\gamma_{G/PL}$ is fiber homotopy trivial so its associated Thom spectrum is just the suspension spectrum of $G/PL_+ = G/PL \cup \{+\}$, and we have

$$M(\gamma_{SO} \times - \gamma_{G/PL}) = MSO \wedge (G/PL_+) .$$

Thus we can identify $\pi_*(M(\gamma_{SO} \times - \gamma_{G/PL}))$ with $\Omega_*(G/PL)$.

---

*) Here $\Omega_*(X)$ denotes the smooth bordism of X, cf. Chapter 1.C.

LEMMA 8.3. *On making the identification above the map* η *is the composite*

$$\Omega_*(G/PL) \longrightarrow \pi_*(M(\gamma_{SO} \times - \gamma_{G/PL})) \xrightarrow{M(\theta)_*} \pi_*(MSPL) = \Omega_*^{PL} .$$

Proof. Let $f: M \to G/PL$ represent an element in $\Omega_*(G/PL)$, then to represent its homotopy class we take $\nu \times f: M \to BSO \times G/PL$ where ν classifies the normal bundle of M. By transversality and the Pontrjagin-Thom construction this is equivalent to

$$S^L \xrightarrow{\ c\ } M(\nu_M) \xrightarrow{M(\nu \times f)} MSO \wedge (G/PL_+) .$$

Now, the normal bundle to $\tilde{M}$ (associated to the map $f: M \to G/PL$) is $\pi^*(\gamma) - \pi^* f^*(\gamma_{G/PL})$. Thus the composite

$$\theta(\nu \times f)\pi : \tilde{M} \to M \to BSO \times G/PL \to BSPL$$

classifies the normal bundle to $\tilde{M}$. On passing to Thom spaces we have the diagram

$$\begin{array}{ccccc}
S^L & \xrightarrow{\ \tilde{c}\ } & M(\nu_{\tilde{M}}) & \xrightarrow{M(\nu_{\tilde{M}})} & MSPL \\
\downarrow{\scriptstyle Id} & & & & \uparrow{\scriptstyle M(\theta)} \\
S^L & \xrightarrow{\ c\ } & M(\nu_M) & \longrightarrow & MSO \wedge (G/PL_+)
\end{array}$$

which is easily seen to be commutative. This completes the proof.

LEMMA 8.4. *The only torsion in* $\Omega_*(G/PL)$ *is 2-torsion. Moreover, the Hurewicz homomorphism* $\Omega_*(G/PL) \to H_*(G/PL; \mathbf{Z})$ *is onto.*

REMARK 8.5. In particular 8.4 implies the Atiyah-Hirzebruch spectral sequence with E_2 term $H_*(G/PL, \Omega_*(pt))$ and converging to $\Omega_*(G/PL)$ has $E_2 = E_\infty$, and also shows

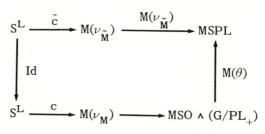

$$\Omega_*(G/PL)/\mathrm{Tor} \cong F_*(G/PL) \otimes \Omega_*(pt)/\mathrm{Tor}\,.$$

where as usual $F_*(G/PL) = H_*(G/PL; \mathbf{Z})/\mathrm{Tor}$.

Proof of 8.4. We check first at the prime 2 where the result is clear from 1.35. At odd primes $\Omega_*(G/PL) \otimes \mathbf{Z}_{(p)} \cong \Omega_*(BO) \otimes \mathbf{Z}_{(p)}$ by 4.28 and so it suffices to check for BO. Consider the map

$$r: BU \to BO$$

which pulls back the universal bundle to the universal complex bundle. We know $H_*(BU) = P\{b_2, b_4, \cdots, b_{2i} \cdots\}$ and at odd primes $H_*(BO) = P\{r_*(b_4),\ r_*(b_8), \cdots, r_*(b_{4i}), \cdots\}$. Moreover, the map

$$H: CP^n \to BU$$

classifying the canonical line bundle satisfies $H_*[CP^n] = b_{2n}$. Thus, $(rH)_*[CP^{2n}] = r_*(b_{4n})$ and

$$r_*(H \times \cdots \times H)_*[CP^{2n_1} \times \cdots \times CP^{2n_r}]$$

$$= r_*(H_*[CP^{2n_1}] \cdots H_*[CP^{2n_r}]) = r_*(b_{4n_1}) \cdots r_*(b_{4n_r})\,.$$

This shows the Hurewicz map is onto at odd primes, implies the triviality of the Atiyah-Hirzebruch spectral sequence, and,since $\Omega_*(pt) \otimes \mathbf{Z}[½] = P\{x_4, \cdots, x_{4n}, \cdots\}$ has no torsion completes the proof of 8.4.

In 1.35 we reviewed Wall's result on the homotopy type of the smooth 2-local Thom spectrum MSO[2]. The corresponding result in the PL category is due to Browder, Liulevicius and Peterson [29]. They prove that MSPL[2] is again a wedge of suspensions of Eilenberg-MacLane spectra. The summands which occur this time are $\Sigma^r K(\mathbf{Z}_{(2)})$ and $\Sigma^r K(\mathbf{Z}/2^s)$ (for all $s > 0$). The torsion free summands $\Sigma^r K(\mathbf{Z}_{(2)})$ are in one-to-one correspondence with an additive basis for $F_*(BSPL) \otimes \mathbf{Z}_{(2)}$, and we have

LEMMA 8.6. *The composition*

$$\Omega_*^{PL}/\mathrm{Tor}\otimes Z_{(2)} \xrightarrow{h} F_*(MSPL)\otimes Z_{(2)} \xrightarrow{\Phi} F_*(BSPL)\otimes Z_{(2)}$$

is an isomorphism. Here h *is the Hurewicz homomorphism and* Φ *the Thom isomorphism.*

THEOREM 8.7. *The map* $\eta: \Omega_*(G/PL)/\mathrm{Tor} \to \Omega_*^{PL}/\mathrm{Tor}$ *is onto.*

Proof. Using the identification in 8.3 the theorem follows from 5.21 at odd primes. At the prime 2 the situation is more complicated. In view of 8.6 we must show that

$$\theta_*: F_*(BSO; Z_{(2)}) \otimes F_*(G/PL; Z_{(2)}) \to F_*(BSPL; Z_{(2)})$$

is surjective. In Chapter 13 we will calculate the Bockstein spectral sequence with initial term $H_*(BTOP; Z/2)$ which converges to $F_*(BTOP)\otimes Z/2$. In particular we will show in Theorem 13.20 that

$$F_*(BSO)\otimes Z/2 \otimes F_*(G/TOP)\otimes Z/2 \xrightarrow{\theta'} F_*(BTOP)\otimes Z/2$$

is onto (with θ' defined analogously to θ above). This implies that

$$\theta'_*: F_*(BSO; Z_{(2)}) \otimes F_*(G/TOP; Z_{(2)}) \to F_*(BTOP; Z_{(2)})$$

is also onto. Finally, note from 4.8 and 4.36 that the natural map $F_*(G/PL; Z_{(2)}) \to F_*(G/TOP; Z_{(2)})$ is an isomorphism. On the other hand $F_*(BPL; Z_{(2)})$ surely injects into $F_*(BTOP; Z_{(2)})$ and since θ'_* is onto, $F_*(BPL; Z_{(2)}) \to F_*(BTOP; Z_{(2)})$ is an isomorphism. This completes the proof.

The last part of the proof of 8.7 also shows

LEMMA 8.8. *The natural map* $\Omega_*^{PL}/\mathrm{Tor} \to \Omega_*^{TOP}/\mathrm{Tor}$ *is an isomorphism.*

REMARK 8.9. The structure of the mod 2 Bockstein spectral sequence of BTOP, used in the proof of 8.7 represents one of the main calculational efforts of the book. We note for any finite r that it is definitely false that

$$\theta'_*: E^r(BSO) \otimes E^r(G/TOP) \to E^r(BTOP)$$

is onto. Thus the (2-primary) torsion structure of Ω_*^{TOP} and Ω_*^{PL} are far more complicated to describe (compare Chapter 14).

We call a PL manifold $\widetilde{CP}^{2n}$ an *exotic projective space* if $\widetilde{CP}^{2n}$ is homotopy equivalent to CP^{2n}. There is precisely one oriented homotopy equivalence $\widetilde{CP}^{2n} \to CP^{2n}$ so $\widetilde{CP}^{2n}$ determines an element of $[CP^{2n}, G/PL]$ usually called its normal invariant (cf. 2.14 and 2.23). On the other hand, any element γ of $[CP^{2n}, G/PL]$ gives by transversality a normal map with range CP^{2n}, and, if the associated surgery obstruction vanishes, an exotic complex projective space. If the surgery obstruction of γ is not zero then, using 2.24 the surgery obstruction of

$$\gamma': CP^{2n} \longrightarrow CP^{2n} \vee S^{2n} \xrightarrow{\gamma \vee (k \cdot \iota_{4n})} G/PL$$

is zero for a suitable integer k. Thus the domain of the surgery problem of γ is of the form $CP^{2n} \# -k \cdot M^{4n}$ where M^{4n} is the Milnor manifold of index 8 (the domain of the surgery problem associated to ι_{4n}, cf. 2.16 and subchapter 8.B below).

THEOREM 8.10. *A set of generators for* $\Omega_*^{PL}/\mathrm{Tor}$ *is contained in the set consisting of the index 8 Milnor manifolds, the differentiable generators and the exotic complex projective spaces.*

Proof. At the prime 2 it is easy to see that $\Omega_*(G/PL)/\mathrm{Tor} \otimes Z_{(2)}$ is generated over $\Omega_*(pt)$ by generating sphere maps $\iota_{4n}: S^{4n} \to G/PL$ and maps $f: CP^{2^{i+1}n} \to G/PL$ which satisfy $f^*(K_{4n}) = e^{2n}$. Indeed, the images of the orientation classes $[\delta^{2n}]$, $[CP^{2^{i+1}n}]$ under these maps form a basis for $F_*(G/PL) \otimes Z_2$. The homomorphism η associates to the sphere map the Milnor manifold of index 8, M^{4n}, and to f a manifold of the form $\widetilde{CP}^{2^{i+1}n} \# - kM^{2^{i+2}n}$.

At odd primes $G/PL[p]$ is H-equivalent to $BO^{\otimes}[p]$ by 4.34 and $BO^{\otimes}[p]$ is H-equivalent to $BO^{\oplus}[p]$. We have seen in the proof of 8.4 that a set of generators for $\Omega_*(BO) \oplus Z_{(p)}$ over $\Omega_*(pt)$ consists of the maps $rH: CP^{2n} \to BO$. Again η associates to such a map a manifold of the form $\widetilde{CP}^{2n} \# - kM^{4n}$, and since η is a ring homomorphism this completes the proof.

We now wish to elucidate 8.10 by constructing models for the torsion free generators.

B. *The Kervaire and Milnor manifolds* (*see also* [27])

We can plumb together two or more copies of the tangent disk bundle to S^n by the following device. Let $D^n \subset S^n$ be a small disk, then $\tau(S^n) \mid D^n$ is trivial so $\tau_{Disk}(S^n) \mid D^n \cong D^n \times D^n_{fiber}$. Then we plumb $\tau_{Disk}(S^n)^1$ with $\tau_{Disk}(S^n)^2$ on identifying $(D^n \times D^n_{fiber})^1$ to $(D^n \times D^n_{fiber})^2$ by setting $(x,y)^1 \sim (y,x)^2$

8.11

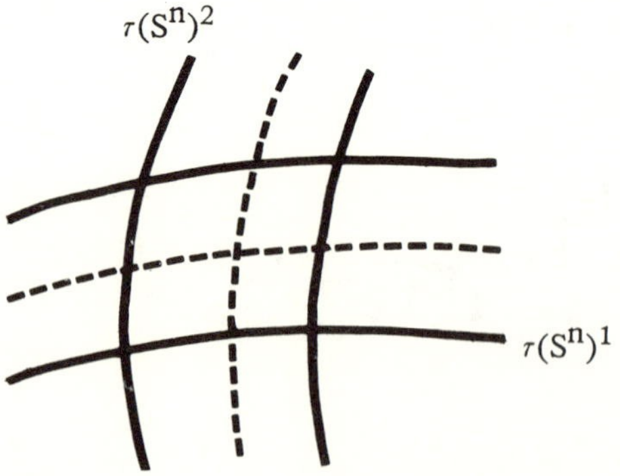

If 2 or more tangent disk bundles are to be plumbed to $\tau_{Disk}(S^n)$ we take disjoint disks $D^n_1 \cdots D^n_r \subset S^n$, one for each bundle to be attached.

The resulting manifold with boundary $(M, \partial M)$ can be concisely described. It has the homotopy type of a wedge of spheres $S^n \vee \cdots \vee S^n$

one for each tangent bundle, and the homology of ∂M which is $n-2$ connected is described in terms of the intersection matrix for M.*) For n even, this matrix has 2's along the diagonal, (for n odd, zeros) and ± 1 in the (i, j) and (j, i) position if the i'th τ is plumbed to the j'th. More precisely, we have the exact sequence

$$8.12 \qquad 0 \longrightarrow H_n(\partial M) \longrightarrow H_n(M) \xrightarrow{A} H_n(M) \longrightarrow H_{n-1}(\partial M) \longrightarrow 0$$

where A is the intersection matrix and we have used duality to identify $H_n(M)$ with $H_n(M, \partial M)$. In particular, $H_{n-1}(\partial M) = 0$ if and only if $\det(A) = \pm 1$, and in this case $H_n(\partial M) = 0$ as well so ∂M has the homology type of S^{2n-1}.

In all cases it is easy to see that M is parallelizable, and if $n \geq 3$ then $\pi_1(\partial M) = \pi_1(M) = 0$, but this is not necessarily true when $n = 2$.

Summarizing

LEMMA 8.13. *If* $\det A = \pm 1$ *and* $n \geq 3$ *then* ∂M^{2n} *is* PL *homeomorphic to* S^{2n-1}.

Proof. $2n-1 \geq 5$, M is a differentiable manifold having the homotopy type of S^{2n-1}, and so by the generalized Poincaré conjecture is PL homeomorphic to S^{2n-1}.

In particular we can construct the closed PL-manifold M_A^{2n} by attaching the cone on ∂M to M.

EXAMPLE 8.14 (The Kervaire manifolds). For n odd $(n \geq 3)$ plumb together 2 copies of $\tau_{\mathrm{Disk}}(S^n)$, then $A = \begin{pmatrix} 0 & 1 \\ -1 & 0 \end{pmatrix}$, $\det(A) = +1$, and the resulting PL manifold M_A^{2n} is called the Kervaire manifold. If

*) Strictly, this is only true if the graph of the plumbings is simply connected (see e.g. 8.15). However, we implicitly assume this throughout the remaining discussion.

$n = 1$, M_A^2 is easily seen to be the torus. However, if $n \geq 3$, then M_A^{2n} is more complex. In particular for $n \neq 2^i - 1$, M_A^{2n} is neither differentiable nor even PL-bordant to a differentiable manifold. (In [35] it is shown that $M_A^{2^i-2}$ is at least PL-bordant to a differentiable manifold. M^{2^i-2} is differentiable if and only if there exists an element in $\pi^S_{2^i-2}(S^0)$ with Arf invariant 1 ([26]). At present this is known only for $i \leq 6$, so M_A^2, M_A^6, M_A^{14}, M_A^{30} and M_A^{62} are differentiable ([17]).)

EXAMPLE 8.15 (The index 8 Milnor manifold). Plumb together 8 copies of $\tau_{\mathrm{Disk}}(S^{2n})$ according to the diagram E_8:

The resulting intersection matrix

$$B = \begin{pmatrix} 2 & 1 & 0 & 0 & 0 & 0 & 0 & 0 \\ 1 & 2 & 1 & 0 & 0 & 0 & 0 & 0 \\ 0 & 1 & 2 & 1 & 1 & 0 & 0 & 0 \\ 0 & 0 & 1 & 2 & 0 & 0 & 0 & 0 \\ 0 & 0 & 1 & 0 & 2 & 1 & 0 & 0 \\ 0 & 0 & 0 & 0 & 1 & 2 & 1 & 0 \\ 0 & 0 & 0 & 0 & 0 & 1 & 2 & 1 \\ 0 & 0 & 0 & 0 & 0 & 0 & 1 & 2 \end{pmatrix}$$

is seen directly to have determinant $+1$ and for $n \geq 2$ the resulting manifold M_B^{4n} is the index 8 Milnor manifold. It is never differentiable nor even PL-bordant to a differentiable manifold [35].

We write

$$mM_B^{4n} = M_B^{4n} \# M_B^{4n} \# \cdots \# M_B^{4n}$$

to denote the connected sum of m copies of M_B^{4n}.

THEOREM 8.16 (Kervaire-Milnor, Quillen). mM_B^{4n} *is differentiable if and only if* m *is a multiple of*

$$a_n 2^{2n-2}(2^{2n-1}-1)\ \mathrm{Num}(B_{2n}/4n)$$

(*see e.g.* [62]).

REMARK 8.17. A beautiful method for representing the M_B^{4n}, M_A^{2n} in terms of algebraic hypersurfaces with isolated singularities has been discovered by E. Brieskorn (see e.g. [22], [99]).

The Kervaire and Milnor manifolds are especially important as they serve to describe normal maps associated to elements in $\pi_*(G/PL)$. This was used for example in the proofs of 2.24 and 8.10. Specifically, the degree 1 map $\pi: M_B^{4n} \to S^{4n}$ is covered by a bundle map $\hat{\pi}: \nu \to \zeta$ to give a degree 1 normal map. The associated map $\iota: S^{4n} \to G/PL$ represents a generator of $\pi_{4n}(G/PL)$, $n > 1$. Similarly for the Kervaire manifolds (cf. 2.16).

C. *Constructing the exotic complex projective spaces*

We now construct the remaining generators for the torsion free parts of the Top and PL oriented bordism rings.

Given a PL normal map $f: \tilde{M} \to M$ and an n-plane bundle ξ over M we obtain a new normal map

$$\tilde{f}: f^*(\xi) \to \xi$$

extending f. Restricting to disk bundles we obtain a degree 1 normal map of manifolds with boundary. Certainly, since $\tilde{f}$ can be thought of as a normal bordism of $\tilde{f}|\partial$ to the trivial normal problem, $\tilde{f}|\partial$ is bordant to a homotopy equivalence. But it may not be possible to find a manifold in the normal cobordism class of $\tilde{f}|\partial$ so $\tilde{f}$ is a PL-homeomorphism, this depends on more subtle questions.

In the special case when $\partial\xi$ is a sphere S^n, $n \geq 5$ then any PL-manifold homotopic to $\partial\xi$ is also PL-homeomorphic to it. In particular we could start with CP^4. There is a degree 1 normal map

8.18 $$f: CP^4 \# mM_B^8 \to CP^4 \# mS^8 = CP^4 .$$

Now, if H_4 is the canonical line bundle over CP^4 then $\partial H_4 = S^9$ and on $\partial f^*(H_4)$ the degree 1 surgery problem is, as observed above, normally bordant to a PL-homeomorphism. Let $g: W^{10} \to I \times S^9$ be the bordism, then with a little care we can choose W^{10} so

8.19 $$h: f^*(H_4) \cup_\partial W^{10} \to H_4$$

becomes a homotopy equivalence. (This is not a general argument, it depends on the special form of 8.18. The idea is to attach the minimum number of handles necessary to kill the surgery kernel on $\partial f^*(H_4)$.)

Now cone off the common boundaries in 8.19 to obtain from $f^*(H_4) \cup_\partial W^{10}$ an exotic CP^5 together with a homotopy equivalence extending h

8.20 $$\bar{h}: \widetilde{CP}^5 \to CP^5 ,$$

that is, a homotopy triangulation $(\widetilde{CP}^5, \bar{h})$ of CP^5. This determines a degree 1 normal map over CP^5 and hence an element $\lambda(\widetilde{CP}^5, \bar{h}) \in [CP^5, G/PL]$.

LEMMA 8.21. *The normal invariant* $\lambda = \lambda(\widetilde{CP}^5, \bar{h})$ *is never trivial. In fact,*

$$\lambda^*(K_8) = m \cdot e^4$$

where e *is the 2-dimensional generator and* K_8 *is the characteristic class from* 4.6.

Proof. By construction $\bar{h}$ is transversal on CP^4 with transversal image $CP^4 \# mM_B^8$; it is also transversal on CP^2 with transversal image CP^2.

Thus $\lambda|CP^2$ is trivial and $\lambda|CP^4$ is the normal invariant of 8.18. But 8.18 has surgery obstruction equal to m and 8.21 follows from 4.9.

Combining 8.21 with 4.22 we can calculate the $\mathcal{L}$-genus of $\widetilde{CP}^5$,

8.22 $$\mathcal{L}(\widetilde{CP}^5) = \mathcal{L}(CP^5)\cdot(1+8me^4)\,.$$

In particular since the rational Pontrjagin classes are topological invariants so are the $\mathcal{L}$-classes and this shows $\widetilde{CP}^5$ is not homeomorphic to CP^5.

We now iterate the construction. Thus in the problem above, we could consider the new normal map

8.23 $$f\colon \widetilde{CP}^5 \# M_A^{10} \to CP^5 \# S^{10} = CP^5\,.$$

Once more we can consider the normal problem $f^*(H_5)\to H_5$, make it a homotopy equivalence on the boundary by doing surgery so as to get a normal bordism W^{12} between $\partial(f^*H_5)$ and S^{11} covering $I\times S^{11}$, and satisfying

$$h\colon f^*(H_5)\cup W^{12} \to H_5$$

is a homotopy equivalence of pairs. (To be explicit, note that the 5 and 6 dimensional surgery kernels are $\mathbf{Z}\oplus\mathbf{Z}$, and the remaining ones are 0. Thus we can embed 2 disjoint copies of $S^5\times D^6$ in $\partial f^*(H_5)$ representing the 2 generators. Attaching handles over these embeddings constructs W^{12}.)

As before we can cone off obtaining a normal map

$$\bar{f}\colon f^*(H_5)\cup W^{12}\cup_{S^{11}} D^{12} \to CP^6$$

which is a homotopy equivalence.

A second way of obtaining an exotic CP^6 from $\widetilde{CP}^5$ is to take $\bar{h}^*(H_5)$, (from 8.20); note that $\partial(\bar{h}^*(H_5)) = S^{11}$ and cone off obtaining $\widetilde{CP}^6 \to CP^6$.*)

*) The universal S^1-bundle over $\widetilde{CP}^5$ is S^{11} with a certain free S^1-action. The join of this action with the standard action on S^1 gives a free action on $S^{13} = S^{11} * S^1$ whose orbit space is precisely $\widetilde{CP}^6$.

LEMMA 8.24. *$\widetilde{CP}^6$, CP^6 and $f^*(H_5) \cup W^{12} \cup_{S^{11}} D^{12}$, are topologically distinct homotopy projective spaces.*

Proof. $\widetilde{CP}^6$ and CP^6 are distinguished using the $\mathcal{L}$ class as in 8.22. Unfortunately, CP^6 and $f^*(H_5) \cup W^{12} \cup D^{12}$ both have the same $\mathcal{L}$ class so to distinguish them we must use a further invariant. In [35], it was shown that there is a primitive class $\kappa_{10} \in H^{10}(BTOP, Z/2)$ with $\hat{j}^*(\kappa_{10}) = K_{10} \in H^{10}(G/PL, Z/2)$ being the Kervaire class. Clearly, by the construction there is a degree 1 normal map

$$h'\colon f^*(H_5) \cup W^{12} \cup D^{12} \to \widetilde{CP}^6$$

with h' transversal on $\widetilde{CP}^5$ and $(h')^{-1}(\widetilde{CP}^5) = CP^5 \# M_A^{10}$. Thus there is an associated map $\rho\colon \widetilde{CP}^6 \to G/PL$ with $\rho^*(\kappa_{10}) = e^5$. Now, as has been noted $\nu(f^*(H_5) \cup W^{12} \cup D^{12}) = h'^*\nu(\widetilde{CP}^6) - h'^*\rho^*(\gamma)$ and thus $\nu^*(\kappa_{10}) = h'^*\kappa_{10}(\nu(\widetilde{CP}^6)) - h'^*(e^5)$. We see that κ_{10} of $\nu(f^*(H_5) \cup W^{12} \cup D^{12})$ and $\kappa_{10}(\widetilde{CP}^6)$ are necessarily distinct. Since κ_{10} is a TOP characteristic class the result follows.

Clearly, we may iterate this process in higher and higher dimensions. Thus, suppose we are given an exotic CP^n, (call it $\widetilde{CP}^n$) and a homotopy equivalence $f\colon \widetilde{CP}^n \to CP^n$.

We begin the construction of a new $\widetilde{\widetilde{CP}}^{n+1}$ by first considering the surgery problems

8.25
$$f\colon \widetilde{CP}^n \# mM_A^{2n} \to CP^n \# S^{2n} = CP^n \quad (n \text{ odd}),$$
$$f\colon \widetilde{CP}^n \# mM_B^{2n} \to CP^n \# S^{2n} = CP^n \quad (n \text{ even})$$

where in the first case $m = 0$ or 1, in the second $m \in Z$ and where $0 \cdot M^{2n}$ is to be interpreted as S^{2n}.

Next, take $f^*(H_n)$ and do surgery to just kill the lowest dimensional surgery kernel on the boundary. This gives us a bordism W^{2n} between $\partial f^*(H_n)$ and S^{2n+1} which satisfies the additional condition that

8.26 $$f^*(H_n) \cup {}_\partial W^{2n} \to H_n$$

is a homotopy equivalence of pairs. Cone off and continue.

THEOREM 8.27. *Any* M^{2n} *homotopy equivalent to* CP^n *for which a homotopy equivalence*

$$f: M \to CP^n$$

can be found so the transverse inverse image of CP^2 *is obtained via the constructions above starting with* CP^3.

Proof. The assumptions imply the map associated to f, $h: CP^n \to G/PL$ is trivial on CP^2. Thus, the map factors through the 5-connected cover $G/PL[6, \cdots, \infty]$. Now, the mapping set $[CP^n, G/PL[6, \cdots, \infty]]$ has a spectral sequence converging to it with

$$E_2^{i,j} = H^i(CP^n, \pi_j(G/PL[6, \cdots, \infty]) .$$

Clearly $E_2 = E_\infty$ since only even dimensions occur in both cohomology and homotopy. Now the proof is a routine counting argument, using induction on n.

REMARK 8.28. It is clear that 8.22 and 8.24 generalize to this general situation to show that all these exotic CP^n's are topologically distinct, except that in 8.24 we only know of the existence of the κ_{4j-2} for j not a power of 2. (Indeed, there is no associated κ_{4j-2} when j is a power of 2.) So, in these dimensions the arguments above do not tell us whether modifying by M_A^{4j-2} and doing the constructions above lead to topologically distinct manifolds. However, note that the exotic CP^n from 8.27 determine distinct homotopy triangulations, i.e. distinct elements of $\mathcal{S}_{PL}(CP^n)$, by the obvious generalization of 8.21. An (oriented) homotopy equivalence of CP^n is homotopic to the identity, so in a homotopy triangulation $(\widetilde{CP}^n, h)$ the map h is redundant. It follows that distinct elements of $\mathcal{S}_{PL}(CP^n)$ fall into distinct homeomorphism types (cf. Sullivan [134]).

REMARK 8.29. The constructions of the exotic CP^n's and related constructions have been studied by several authors (e.g. [43], [141]). Note in particular that Wall's construction of all normal invariants of Lens spaces (giving exotic Lens spaces) is essentially the same procedure, but generalized to the non-simply connected case. (Cf. [141], p. 213.)

REMARK 8.30. We may break the procedure above into steps in another way. Let $\widetilde{CP}{}^n_0$ be $\widetilde{CP}{}^n - D^{2n}$. Over $\widetilde{CP}{}^n_0$ we induce $f^*(H_n)$. The boundary of the resulting disk bundle now consists of two parts, the sphere bundle over $\widetilde{CP}{}^n_0$ which is the homotopy type of $D^2 \times S^{2n-1}$ and the disk bundle over $\partial(\widetilde{CP}{}^n_0)$ which is again $S^{2n-1} \times D^2$.

We can write $\widetilde{CP}{}^n \# M = \widetilde{CP}{}^n_0 \cup_\partial M_0$, and note that $f^*(H_n) \mid M_0$ is trivial. Moreover, and this is a key point, the handles needed to construct W can be attached by maps of spheres into the part of the S^1 bundle in $f^*(H_n)$ lying over M_0.

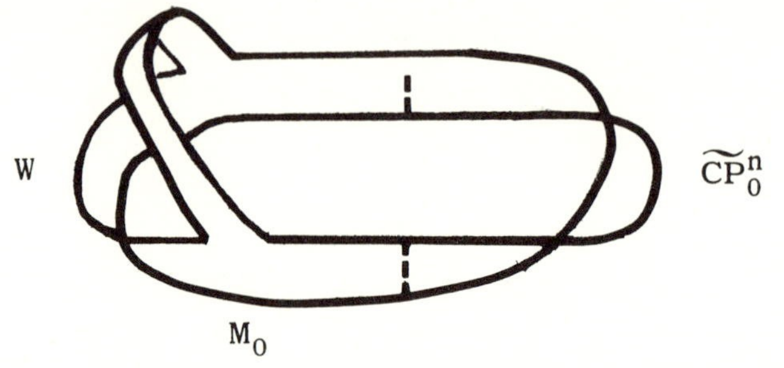

After these handles have been attached we may regard the resulting manifold as

$$(D^2 \times M_0 \cup W) \cup_{D^2 \times S^{2n-1}} (f^*(H_n) \mid \widetilde{CP}{}^n_0) .$$

Now, the space $D^2 \times M_0 \cup W$ is simply connected, has trivial homology and dimension ≥ 6, hence is PL-homeomorphic to a disk. Thus, the manifold above (which is another way of writing 8.26) is obtained by attaching a disk D^{2n+2} to $f^*(H_n) \mid \widetilde{CP}{}^n_0$, by specifying an embedding of $S^{2n-1} \times D^2$ in $\partial D^{2n+2} = S^{2n+1}$, and the different homeomorphism types that result are due to the different knot types of the embeddings.

Specifically, we can take the Brieskorn varieties

$$V_{n,m} = \{z_1^{6m-1} + z_2^3 + z_3^2 + \cdots + z_{2n+1}^2 = \varepsilon\} \quad \text{for} \quad \varepsilon > 0$$

sufficiently small and intersect with a small disk D^{4n+2}. Then $\nu(V_{n,m})$ is trivial and intersecting with S^{4n+1} we get $\Sigma^{4n-1} \times D^2$, where, from [22] Σ is an exotic sphere. Then, identifying this with $S^{4n-1} \times D^2$ gives the desired attaching above replacing 8.26. In the Arf invariant dimensions we can use

$$\{z_1^3 + z_2^2 + \cdots + z_{2n}^2 = \varepsilon\} \cap S^{4n-1} .$$

The CP^n's described above are on the one hand redundant to describe the torsion free generators of $\Omega_*^{PL}(pt)$ (we do not need to use the Kervaire invariant construction) and insufficient. In order to get sufficient manifolds at odd primes we must modify CP^2 as well. This provides no real difficulty, though. Let N^4 be an almost parallelizable manifold of index 16, for example the Kummer surface.

This is given explicitly by blowing up singular points on the quotient of $T^4 = (S^1)^4$ by the action of $Z/2$ acting as $T(z_1, \cdots, z_4) = (\bar{z}_1, \cdots, \bar{z}_4)$. In a neighborhood of each of the 16 singular points $T^4/Z/2$ is a cone on RP^3. Delete the interior of the cones and replace by copies of $\tau_{Disc}(S^2)$, whose boundary is also RP^3, glued along the boundaries. It may be checked that the resulting manifold N^4 is simply connected and that $H_2(N^4; Z)$ is 22 copies of Z. Somewhat less evident is the result that the generators may be chosen as embedded S^2's with normal bundle $\tau(S^2)$. The intersection pairing is rather complicated, however, in this basis. See [105] for details.

There is a degree 1 normal map $h: N^4 \to S^4$ and we may start our construction using

$$f: CP^2 \# kN^4 \to CP^2 .$$

When we have added these resulting spaces we have

THEOREM 8.31. *The constructions above using only* N^4, M_B^{4n} *give us sufficient exotic* CP^n*'s which together with the* M_B^{4n} *and the differentiable generators contain a set of generators for the torsion free parts of* Ω_*^{PL} *and* Ω_*^{TOP}.

CHAPTER 9

THE TORSION FREE COHOMOLOGY OF G/TOP AND G/PL

We are going to use 4.28 to study the Hopf algebra $F^*(G/TOP) = F^*(G/PL)$ and in the next chapters we apply 5.12 to get essentially complete information on $F^*(BTOP) = F^*(BPL)$. In both cases we will need to have a solid grip on the Hopf algebras $F^*(BSO^{\oplus})$ and $F^*(BSO^{\otimes})$.

A. *An important Hopf algebra*

Consider the graded Hopf algebra $H_d(A)$ over a unitary subring A of $\mathbf{Q}$.

9.1
$$H_d(A) = P\{h_1, h_2, \cdots\}, \quad \deg(h_i) = 2id$$
$$\psi(h_n) = \sum_{i=0}^{n} h_i \otimes h_{n-i}, \qquad h_0 = 1$$

which occurs in particular as $F_*(BSO^{\oplus}) \otimes A$ for $d = 2$.

Let $h(x)$ be the power series

$$h(x) = \sum_{i=0}^{\infty} (-1)^i h_i x^i$$

and define homogeneous polynomials $s_n = s_n(h_1, \cdots, h_n)$ by

9.2
$$-x \frac{d}{dx} \log h(x) = \sum_{n=1}^{\infty} s_n x^n .$$

The elements s_n (in degree 2nd) are the Newton polynomials. Specifically, we view h_n as the n'th elementary symmetric function in the variables $t_1, t_2, \cdots$, so formally we can write

$$h(x) = \Pi(1 - t_i x) \ ,$$

then s_n becomes the polynomial associated with $\Sigma\, t_i^n$. Alternately, the s_n can be specified by the recursion formula

$$9.3 \qquad s_n - s_{n-1}h_1 + \cdots + (-1)^{n-1}s_1 h_{n-1} + (-1)^n n h_n = 0 \ .$$

LEMMA 9.4. *In each degree* 2nd *the module of primitive elements is a single copy of* A *generated by* s_n.

Proof. In degrees less than 2dN we can consider $H_d(A)$ as the subring of invariant elements in $P\{t_1, \cdots, t_N\}$ under the action of the permutation group Σ_N. Let $t'_1, \cdots, t'_N$ be another set of indeterminates. Then

$$\sigma_n(t_1, \cdots, t_N, t'_1, \cdots, t'_N) = \sum_{i=0}^{n} \sigma_i(t_1, \cdots, t_N) \cdot \sigma_{n-i}(t'_1, \cdots, t'_N) \ ,$$

where σ_j denotes the j'th elementary symmetric function, and we have the commutative diagram

$$\begin{array}{ccc} H_d(A) & \xrightarrow{\ \psi\ } & H_d(A) \otimes H_d(A) \\ \Big\downarrow{\scriptstyle i_{2N}} & & \Big\downarrow{\scriptstyle i_N \otimes i'_N} \\ P\{t_1, \cdots, t_N, t'_1, \cdots, t'_N\}^{\Sigma_{2N}} & \hookrightarrow & P\{t_1, \cdots, t_N\}^{\Sigma_N} \otimes P\{t'_1, \cdots, t'_N\}^{\Sigma_N} \ . \end{array}$$

It follows that $\psi(x) = x \otimes 1 + 1 \otimes x$ if and only if $i_{2N}(x) = i_N(x) + i'_N(x)$. But the only elements in $P\{t_1, \cdots, t_N, t'_1, \cdots, t'_N\}^{\Sigma_{2N}}$ with this property are of the form $a(\Sigma t_i^n + \Sigma t_i'^n)$, $a \in A$. This proves the lemma.

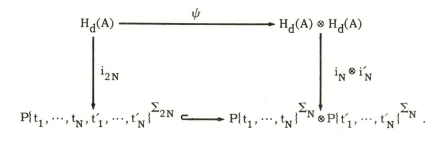

Later in this chapter we are going to present various ways of constructing a generating genus for Hopf algebras of the type $H_d(A)$. It is therefore useful to know the group $\mathrm{Aut}(H_d(A))$ of Hopf algebra automorphisms of $H_d(A)$. If $f \in \mathrm{Aut}(H_d(A))$ then f will map primitive elements to primitive elements, $f(s_n) = u_n s_n$, where u_n is a unit in A. If $A = Q$ then each sequence $\{u_n\}$ can be realized this way since the Newton elements s_n are algebra generators for $H_d(Q)$. For an arbitrary unitary subring A of Q, $H_d(A)$ is torsion free and $\mathrm{Aut}(H_d(A))$ must be a subgroup of $\prod_1^\infty A^\times$. In the basic case $A = \mathbf{Z}$, A. Liulevicius pointed out to us the following simple result

LEMMA 9.5. $\mathrm{Aut}(H_d(\mathbf{Z})) \cong \mathbf{Z}/2 \oplus \mathbf{Z}/2$.

Proof. Let $f: H_d(\mathbf{Z}) \to H_d(\mathbf{Z})$ be a Hopf algebra automorphism with $f(s_1) = s_1$ and $f(s_2) = s_2$. We show that $f(h_n) = h_n$ for all n. Indeed, assume inductively that $f(h_i) = h_i$ for $i < n$. Then $f(h_n) - h_n = \lambda\, s_n$ and the coefficient of h_n in $f(h_n)$ is $1 - (-1)^n \lambda n$. This must be ± 1 which for $n > 2$ is only possible if $\lambda = 0$. But $f(h_1) = h_1$ and $f(h_2) = h_2$ by assumption and $f = \mathrm{Id}$.

On the other hand the canonical (anti) automorphism χ defined by

$$\chi(g) + g + \Sigma\chi(g'_i)\cdot g''_i = 0 \quad \text{where} \quad \psi(g) = 1 \otimes g + g \otimes 1 + \Sigma g'_i \otimes g''_i$$

maps s_n to $-s_n$. The automorphism $\Phi, \Phi(h_n) = (-1)^n h_n$ sends s_1 to $-s_1$ and s_2 to s_2. Hence $\mathrm{Aut}(H_d(\mathbf{Z})) = \mathbf{Z}/2 \oplus \mathbf{Z}/2$ with generators Φ and χ.

The Hopf algebra $H_d(A)$ is self dual. Indeed let $p_n \in H_d(A)^*$ be the element given by

$$\text{9.6} \qquad \begin{aligned} &\langle p_n, h_1^n \rangle = 1 \\ &\langle p_n, h_1^{i_1} \cdot \cdots \cdot h_n^{i_n} \rangle = 0 \quad \text{otherwise} \end{aligned}$$

where $<,>$ denotes the evaluation pairing between $H_d(A)^*$ and $H_d(A)$. Then p_n is indecomposable since h_1^n occurs in s_n with coefficient 1, and it is directly seen that $\psi(p_n) = \Sigma p_i \otimes p_{n-i}$. It follows that

$$H_d(A)^* = P\{p_1, p_2, \cdots\}, \quad \psi(p_n) = \sum_{i=0}^{n} p_i \otimes p_{n-i}.$$

(In topology this situation occurs with $H_d(Z) = F_*(BSO)$, h_n = the generator of Image$(F_{4n}(BSO_2)) \to F_{4n}(BSO))$ and p_n the n'th Pontrjagin class, cf. Chapter 1.B.)

We next review results of Husemoller [58] on the structure of $H_d(Z_{(p)})$.

First, note that $H_d(Q)$ splits as a tensor product of monogenic Hopf algebras,

$$H_d(Q) = \bigotimes_{m=1}^{\infty} P\{s_m\}.$$

Indeed, the recursion formula 9.3 shows that the generators h_n can be expressed as polynomials in the s_m, $h_n = -(-1)^n s_n/n + \cdots$. Formally, 9.2 may be rewritten as

$$\sum (-1)^i h_i x^i = \exp\left(-s_1 x - \frac{s_2 x^2}{2} - \frac{s_3 x^3}{3} - \cdots\right).$$

In order to get a splitting of $H_d(Z_{(p)})$ we must replace the series on the right by a series with coefficients in $Z_{(p)}$. To this end we consider the Artin-Hasse exponential series

$$L(1-x) = \exp(-x - x^p/p - \cdots - x^{p^i}/p^i - \cdots).$$

Let $\mu(n)$ denote the Möbius function: $\mu(1) = 1$, $\mu(n) = 0$ if n is divisible by a square and $\mu(p_1 \cdots p_r) = (-1)^r$ if $p_1, \cdots, p_r$ are distinct primes. Then we have

9.7 $$\log L(1-x) = \prod_{(n,p)=1} (1-x^n)^{\mu(n)/n}.$$

Indeed, $\sum_{n|\ell} \mu(n) = 0$ if $\ell > 1$ so

$$L(1-x) = \sum_{(d,p)=1} \frac{1}{\ell} \log L(1-t^{\ell}) \cdot \sum_{n|d} \mu(n)$$

$$= \sum_{(n,p)=1} \sum_{(m,p)=1} \frac{\mu(n)}{nm} \log L(1-x^{mn})$$

$$= \sum_{(n,p)=1} \frac{\mu(n)}{n} \sum_{(m,p)=1} \left(-\frac{x^{nm}}{m} - \frac{x^{nmp}}{mp} - \cdots - \frac{x^{nmp^i}}{mp^i}\right)$$

$$= \sum_{(n,p)=1} \frac{\mu(n)}{n} \log(1-x^n) .$$

Note from 9.7 that $L(1-x)$ has $Z_{(p)}$ integral coefficients. Hence we can define elements $h_{n,i} \in H_d(Z_{(p)})$ for $(n,p) = 1$ and $i \geq 0$ of degree $2p^i nd$ by

9.8
$$\sum_{i=0}^{\infty} (-1)^i h_i x^i = \prod_{(n,p)=1} \prod_{i=0}^{\infty} L(1-h_{n,i}\, x^{p^i n})^{1/n} .$$

It is straightforward to see that $h_{p^i n} \equiv -((-1)^{p^i n}/n)\, h_{n,i} + \cdots$ so $\{h_{n,i}\}$ is a new system of algebra generators for $H_d(Z_{(p)})$, which we call the *Witt vector basis*. The main advantage of this basis is that the primitive elements s_m of $H_d(Z_{(p)})$ admit a very simple expression in terms of the $h_{n,i}$,

9.9
$$s_m = p^i h_{n,i} + p^{i-1} h^p_{n,i-1} + \cdots + h^{p^i}_{n,0}, \quad m = p^i n$$

(apply $-x \frac{d}{dx} \log$ to 9.8).

For each n prime to p we define

$$W_{d,n}(Z_{(p)}) = P\{h_{n,0}, h_{n,1}, \ldots, h_{n,i}, \ldots\} .$$

This is a subalgebra of $H_d(Z_{(p)})$ and in fact a sub Hopf algebra (this is obvious over Q since $H_d(Q)$ is primitively generated but then it follows over $Z_{(p)}$ as $H_d(Z_{(p)})$ is torsion free). We have proved,

THEOREM 9.10 (Husemoller). *$H_d(Z_{(p)})$ is Hopf algebra isomorphic to* $\bigotimes_{(n,p)=1} W_{d,n}(Z_{(p)})$.

REMARK. There are similar splitting results for $H_d(Z[1/p_1, 1/p_2, \cdots])$ where $p_1, p_2, \cdots$ is any set of primes. E.g. for $H_d(Z[\frac{1}{2}])$ one uses the power series

$$\begin{aligned} G(1-x) &= \exp(-x - x^3/3 - \cdots - x^{2n+1}/2n+1 - \cdots) \\ &= (1-x)(1-x^2)^{-\frac{1}{2}} \end{aligned}$$

and defines new generators $k_{2^j, 2q+1}$ of degree $2^{j+1}(2q+1)d$ by

$$\Sigma(-1)^i h_i x^i = \prod_{j=0}^{\infty} \prod_{q=0}^{\infty} G(1-k_{2^j, 2q+1} x^{2^j(2q+1)})^{\frac{1}{2}} .$$

Then

$$W_{d,j}(Z[\tfrac{1}{2}]) = P\{k_{2^j,1}, k_{2^j,3}, \cdots\}$$

is a sub Hopf algebra of $H_d(Z[\frac{1}{2}])$ and $H_d(Z[\frac{1}{2}]) = \bigotimes_{j=0}^{\infty} W_{d,j}(Z[\frac{1}{2}])$.

The convenience of the Witt vector basis is illustrated in our next result—*the integrality lemma*. Let $\mathfrak{A} = 1 + A_1 + A_2 + \cdots$ be a genus in the dual Hopf algebra $H_d(Q)^*$. (A_n is an indecomposable element in degree 2nd and $\psi(A_n) = \Sigma A_i \otimes A_{n-i}$ (cf. Chapter 1.E).) Let $p_1, p_2, \cdots$ be the (integral) generators of $H_d(Z)^*$ constructed in 9.6 and let

$$P(\mathfrak{A}) = \sum_{n=0}^{\infty} (-1)^n \pi_n z^n \in Q[[z]]$$

be the primitive series associated to $\mathfrak{A}$: π_n is the coefficient of p_n in A_n (cf. 1.37).

LEMMA 9.11. *A necessary and sufficient condition that* $\mathfrak{A}$ *be* $Z_{(p)}$*-integral* ($\mathfrak{A} \in H_d(Z_{(p)})^*$) *is that*

(i) *the coefficients* π_n *in* $P(\mathfrak{A})$ *belong to* $\mathbf{Z}_{(p)}$.

(ii) $\pi_{p^i n} \equiv \pi_{p^{i-1} n} \pmod{p^i}$.

Proof. Let $\{h_{n,i}\}$ be the Witt vector basis in $H_d(\mathbf{Z}_{(p)})$ constructed in 9.8. Then $\mathfrak{A}$ is $\mathbf{Z}_{(p)}$ integral if and only if $<\mathfrak{A}, h_{n,i}> \in \mathbf{Z}_{(p)}$ for all $h_{n,i}$. For $i = 0$, we obtain (i) since the coefficient of h_1^n in $s_n(=h_{n,0})$ is 1. But, for example with n prime to p

$$<\mathfrak{A}, s_{pn}> = <\mathfrak{A}, p\, h_{n,1}> + <\mathfrak{A}, h_{n,0}>^p$$

or

$$<\mathfrak{A}, s_{pn}> \equiv <\mathfrak{A}, s_n> \qquad (\text{mod } p)$$

since $\lambda^p \equiv \lambda (\text{mod } p)$ for $\lambda \in \mathbf{Z}_{(p)}$. In general for $\lambda \in \mathbf{Z}_{(p)}$, $\lambda^{p^i} \equiv \lambda^{p^{i-1}} (\text{mod } p^i)$ so that

$$\begin{aligned} <\mathfrak{A}, s_{p^i n}> &= p^i <\mathfrak{A}, h_{n,i}> + \cdots + <\mathfrak{A}, h_{n,0}>^{p^i} \\ &\equiv <\mathfrak{A}, s_{p^{i-1} n}> \end{aligned}$$

and the result follows.

Finally, we recall a striking result due to Ravenel and Wilson [114] on bipolynomial Hopf algebras (where H is a bipolynomial Hopf algebra if both H and its dual are polynomial algebras), which will be useful later.

THEOREM 9.12. *If a bipolynomial Hopf algebra over* A *is isomorphic to* $H_d(A)$ *as an algebra then it is isomorphic to* $H_d(A)$ *as a Hopf algebra, where* $A = \mathbf{Z}$, $\mathbf{Z}_{(p)}$ (*or* $\mathbf{Z}[1/2]$).

B. *The Hopf algebras* $F^*(BSO^{\otimes})$ *and* $F^*(G/PL) \otimes \mathbf{Z}[1/2]$

At each prime separately Atiyah and Segal in [14] exhibited an H-equivalence

$$\delta_p : BSO^{\oplus}[p] \to BSO^{\otimes}[p]$$

whose properties will be basic to our further discussion; we recall its definition.

There are natural operations $(k \geq 0)$

$$\lambda^k : KO(X) \to KO(X)$$
$$\gamma^k : KO(X) \to KO(X)$$

where, on an actual vector bundle, λ^k is the k'th exterior power operation and $\gamma^k(x) = \lambda^k(x+k-1)$ ([9], [12]). If $k = 0$ $\lambda^k(x) = 1$, $\gamma^k(x) = 1$ and $\gamma^1(x) = \lambda^1(x) = x$. The associated "total" operations

$$\lambda_t(x) = \sum_{k=0}^{\infty} \lambda^k(x)\, t^k , \qquad \gamma_t(x) = \sum_{k=0}^{\infty} \gamma^k(x)\, t^k$$

give exponential maps

$$\gamma_t, \lambda_t : KO(X) \to KO(X)\,[[t]]$$

and they are related by the formulae $\gamma_t = \lambda_{t/1-t}$ and $\lambda_s = \gamma_{s/1+s}$. If X is a 4n-dimensional complex then γ^k vanishes identically on $\widetilde{KO}(X)$ when $k > 4n$. In particular for each $t \in \mathbf{Z}$ we have an exponential mapping

$$\gamma_t : \widetilde{KO}(X) \to 1 + \widetilde{KO}(X)$$

where X is finite dimensional, and we get a *unique* induced H-mapping

$$\gamma_t : BSO^{\oplus} \to BSO^{\otimes}$$

(compare 4.29).

The Adams operations correspond to the Newton polynomials in the λ^k,

$$\psi^k(x) - \psi^{k-1}(x)\lambda^1(x) + \cdots + (-1)^k k\lambda^k(x) = 0 .$$

Now, it is well known that

$$\psi^k : \widetilde{KO}(S^{4r}) \to \widetilde{KO}(S^{4r})$$

is multiplication with k^{2r} so λ^k is multiplication by $(-1)^{k-1}k^{2r-1}$. Since $\gamma^k(x) = \lambda^k(x+k-1)$ we see that

$$\gamma^k: \widetilde{KO}(S^{4r}) \to \widetilde{KO}(S^{4r})$$

is multiplication by

9.13 $$c(k,r) = \sum_{j=1}^{k} (-1)^{j-1} j^{2r-1} \binom{k-1}{k-j}.^{*)}$$

LEMMA 9.14 (Atiyah-Segal). *Let* p *be an odd prime. For each* $1 \le i \le \frac{p-1}{2}$ *there exists a* $t = t(i)$ *such that*

$$(\gamma_t)_*: \pi_{4r}(BSO^{\oplus}) \otimes Z_{(p)} \to \pi_{4r}(BSO^{\otimes}) \otimes Z_{(p)}$$

is an isomorphism whenever $r \equiv i \left(\text{mod } \frac{p-1}{2}\right)$.

Proof. It suffices to pick a number t so that

$$\gamma_t: \widetilde{KO}(S^{4r}) \otimes Z/p \to 1 + \widetilde{KO}(S^{4r}) \otimes Z/p$$

is an isomorphism when $r \equiv i \left(\text{mod } \frac{p-1}{2}\right)$. Let x_{4r} be the generator of $\widetilde{KO}(S^{4r})$. Then

$$\gamma_t(x_{4r}) = 1 + \left(\sum_{k=1}^{2r} c(k,r)\, t^k \cdot x_{4r}\right)$$

and we must show that $\sum_{k=1}^{2r} c(k,r)\, t^k$ is not identically zero in Z/p. But from 9.13 we have that $c(k,r) = c(k,i)$ in Z/p. Hence in $Z/p[t]$,

*) $c(k,r) = 0$ if $k > 2r$. This follows e.g. by using the realification $r: \tilde{K}(S^{4r}) \to \widetilde{KO}(S^{4r})$. It commutes with γ^k (since it commutes with ψ^k). The generator of $\tilde{K}(S^{4r})$ is represented by the stable class of a 2r-dimensional complex bundle, so γ^k vanishes on $\tilde{K}(S^{4r})$ for $k > 2r$.

$$\sum_{k=1}^{2r} c(k,r)\,t^k = \sum_{k=1}^{2i} c(k,i)\,t^k$$

is a polynomial of degree less than p. It is not constant as $c(1,i) = 1$ and consequently there is a $t(i)$ so $\sum c(k,i)\,t(i)^k \neq 0$ in $\mathbf{Z}/p$. Any $t \in \mathbf{Z}$ representing this $t(i)$ will do in 9.14.

For p an odd prime we have the H-splitting (cf. 5.9)

$$BSO^{\oplus}[p] \simeq BSO_{(1)} \times \cdots \times BSO_{(m)}, \qquad m = \frac{p-1}{2}$$

where $\pi_{4r}(BSO_{(i)}) = \mathbf{Z}_{(p)}$ when $r \equiv i (\mathrm{mod}\ m)$ and $\pi_{4r}(BSO_{(i)}) = 0$ otherwise. Let $e_i : BSO^{\oplus}[p] \to BSO^{\oplus}[p]$ be the idempotent associated to $BSO_{(i)}$, i.e. e_i is the composition of the projection onto $BSO_{(i)}$ and the inclusion $BSO_{(i)} \subset BSO[p]$. Then

9.15 $$\delta_p = \prod_{i=1}^{m} \gamma_{t(i)} \circ e_i : BSO^{\oplus}[p] \to BSO^{\otimes}[p]$$

is the H-eqüivalence constructed by Atiyah and Segal.

If $p = 2$ we must proceed a little differently. The coefficient $c(k,r)$ is zero if $k > 2$ so the polynomial $\sum_{k=1}^{2r} c(k,r)\,t^k$ becomes $t + t^2$, and

$$\gamma_t : \widetilde{KO}(S^{4r}) \otimes \mathbf{Z}/2 \to 1 + \widetilde{KO}(S^{4r}) \otimes \mathbf{Z}/2$$

is multiplication by $t + t^2$ which is zero in $\mathbf{Z}/2$ for every $t \in \mathbf{Z}$. Now, adjoin to $\widetilde{KO}(X)$ a root α of $t^2 - t + 1/3 = 0$. Then

$$\gamma_\alpha : \widetilde{KO}(X) \to 1 + \widetilde{KO}(x)[\alpha]\,.$$

Since $\gamma_t(x) = \gamma_{1-t}(x)$ on the subgroup $\widetilde{KSO}(X)$ and the conjugate of α is $1-\alpha$ one has

$$\gamma_\alpha : \widetilde{KSO}(X) \to 1 + \widetilde{KSO}(X) \subset 1 + \widetilde{KO}(X)[\alpha]$$

and the argument above shows that γ_α induces an isomorphism on $\widetilde{KSO}(S^{4r})$ when tensored with $\mathbf{Z}/2$. Hence

9.16 $$\gamma_\alpha : BSO^{\oplus}[2] \to BSO^{\otimes}[2]$$

is an H-equivalence.

REMARK 9.17. The map γ_α is equal to the map ρ_A^3 considered in Chapter 5.B. Indeed, let L be a complex line bundle and $\bar{L}$ its conjugate. Then $\gamma_\alpha(L+\bar{L}-2) = (1+\alpha(L-1))(1+\alpha(\bar{L}-1)) = 1/3(L+\bar{L}+1)$ so $ph(\gamma_\alpha)$ is the genus with characteristic formal power series

$$1/3(e^{\sqrt{z}}+e^{-\sqrt{z}}+1) = \frac{e^{3/2\sqrt{z}}-e^{-3/2\sqrt{z}}}{3(e^{½\sqrt{z}}-e^{-½\sqrt{z}})}.$$

Hence $ph(\gamma_\alpha) = ph(\rho_A^3)$ (cf. p. 104) and 4.29 implies that $\gamma_\alpha = \rho_A^3$ as claimed.

The total Pontrjagin character $ph = 1 + ph_4 + ph_8 + \cdots$ is multiplicative and hence represents a genus for $F^*(BSO^{\otimes}; Q)$.

COROLLARY 9.18. (i) $F^*(BSO^{\otimes}) = P\{q_4, q_8, \cdots, q_{4n}, \cdots\}$ *where* $\mathcal{Q} = 1+q_4+q_8+\cdots$ *is a genus*

(ii) *For any* $\mathcal{Q}$ *in* (i) *one has for each* n

$$s_n(q_4, \cdots, q_{4n}) = \pm(2n-1)!\,s_n(ph_4, \cdots, ph_{4n}).$$

Proof. For each prime p, $F^*(BSO^{\otimes}) \otimes \mathbf{Z}_{(p)}$ is a bipolynomial Hopf algebra according to 9.15 and 9.16. Hence $F^*(BSO^{\otimes})$ is a bipolynomial Hopf algebra and (i) follows from 9.12. To prove (ii) we evaluate both sides on a homotopy generator $\iota_{4n} \in \pi_{4n}(BSO)$. First,

$$\langle ph_{4n}, \iota_{4n}\rangle = \langle ch_{2n}, c_*(\iota_{4n})\rangle = a_n$$

where c is the complexification and a_n is 1 for n even and 2 for n odd. Hence $\langle (2n-1)!s_n(ph_4, \cdots, ph_{4n}), \iota_{4n}\rangle = (-1)^{n-1}\frac{(2n)!}{2}a_n$. Second, again from 9.15 and 9.16 it follows for each prime p that

$$\delta_p^*(s_n(q_4, \cdots, q_{4n})) = \lambda_p(n)\,s_n(p_4, \cdots, p_{4n})$$

where $\lambda_p(n)$ is a p-local unit and p_{4i} is the i'th Pontrjagin class. But

$$s_n(p_4, \cdots, p_{4n}) = \frac{(2n)!}{2} ph_{4n}$$

and we see that

$$s_n(q_4, \cdots, q_{4n}) = \lambda(n)(2n-1)!s_n(ph_4, \cdots, ph_{4n})$$

where $\lambda(n)$ is a p-local unit for each p, hence $\lambda n = \pm 1$.

According to 9.5 the genus $\mathcal{Q}$ above is specified by s_1 and s_2, that is, there is a unique generating genus $\mathcal{Q} \in F^*(BSO^{\otimes})$ so that $s_1(\mathcal{Q}) = s_1(ph)$, $s_2(\mathcal{Q}) = 3!s_2(ph)$ or equivalently

$$q_4 = p_4 \quad \text{and} \quad q_8 = -(p_8 + 2p_4^2) .$$

Sign change on the primitive elements is a complicated operation on the associated (rational) genus, and does not preserve integrality. For our applications it is important to know the rational reduction of a generating genus $\mathcal{Q}$ for $F^*(BSO^{\otimes})$. We have the following strengthening of 9.18.

THEOREM 9.19. *The genus* $\mathcal{Q} \in F^*(BSO^{\otimes}; \mathbf{Q})$ *defined by*

$$s_n(q_4, \cdots, q_{4n}) = (2n-1)!s_n(ph_4, \cdots, ph_{4n})$$

is integral and $F^*(BSO^{\otimes}) = P\{q_4, q_8, \cdots, q_{4n}, \cdots\}$.

Theorem 9.19 is a statement about polynomial rings: solving the equation one finds q_{4n} as a rational polynomial in the Pontrjagin classes and we claim that the coefficients in this polynomial are integers. Our proof, however is based on 9.15, 9.16 and properties of $F^*(BSO^{\oplus})$.

Proof. It is sufficient to check that $\gamma_t^*(\mathcal{Q})$ is $\mathbf{Z}_{(p)}$ integral for $t \in \mathbf{Z}$ and that $\gamma_\alpha^*(\mathcal{Q})$ is $\mathbf{Z}_{(2)}$ integral. We calculate the primitive series of $\gamma_t^*(\mathcal{Q})$ or equivalently the coefficient of p_{4r} in $\gamma_t^*(q_{4r})$. Let

$\iota_{4r} \in \pi_{4r}(BSO)$ be the standard generator, then

$$<p_{4r}, \iota_{4r}> = (-1)^{r-1}(2r-1)!a_r$$

and

$$\begin{aligned} <\gamma_t^*(\mathcal{Q}), \iota_{4r}> &= <\mathcal{Q}, \gamma_{t*}(\iota_{4r})> \\ &= \frac{(-1)^{r-1}}{r} <s_r(q_4, \cdots, q_{4r}), \gamma_{t*}(\iota_{4r})> \\ &= (2r-1)! \frac{(-1)^{r-1}}{r} <s_r(ph_4, \cdots, ph_{4r}), \gamma_{t*}(\iota_{4r})> \\ &= (2r-1)! <ph_{4r}, \gamma_{t*}(\iota_{4r})> \\ &= (2r-1)!\, a_r \sum_{k=1}^{2r} c(k, r)\, t^k . \end{aligned}$$

We can then read off the coefficient π_r of p_{4r} in $\gamma_t^*(\mathcal{Q})$,

$$\pi_r = (-1)^{r-1} \sum_{k=1}^{2r} c(k, r)\, t^k .$$

From 9.13 it is easily seen that

$$c(k, r p^i) \equiv c(k, r p^{i-1}) \pmod{p^i} ,$$

thus $\pi_{rp^i} \equiv \pi_{rp^{i-1}}$ and 9.11 applies.

The argument that $\gamma_\alpha^*(\mathcal{Q})$ is a $\mathbf{Z}_{(2)}$-integral genus is essentially the same; we leave the details to the reader.

REMARK 9.20. Consider the H-map $\rho_A^k : BSO^{\oplus} \to BSO^{\otimes}[1/k]$ also used in Chapter 5.B. From [2, II, p. 166] we get that ρ_A^k induces multiplication by $b_r = (k^{2r}-1) B_{2r}/4rk^{2r}$ on $\pi_{4r}(BSO)$ where B_{2r} is the 2r'th Bernoulli number (with the convention that $B_{2i+1} = 0$ for $i > 0$, cf. 11.11 below). If $(k, p) = 1$ then $(\rho_A^k)^*(\mathcal{Q})$ is $\mathbf{Z}_{(p)}$ integral and we have from 9.11

$$b_{np^i} \equiv b_{np^{i-1}} \pmod{p^i}.$$

Suppose $2n \not\equiv 0\,(p-1)$ and that k generates $(\mathbf{Z}/p^2)^\times$ and hence $(\mathbf{Z}/p^i)^\times$ for all i. Then $k^{2np^{i-1}} \not\equiv 1 \pmod{p^i}$ and

$$(k^{2np^i}-1)/(k^{2np^{i-1}}-1) \equiv 1 \pmod{p^i},\quad k^{2np^i}/k^{2np^{i-1}} \equiv 1 \pmod{p^i},$$

so we get

$$B_{2np^i}/2np^i \equiv B_{2np^{i-1}}/2np^{i-1} \pmod{p^i}.$$

These are the so-called Kummer congruences (see e.g. [21]).

The genus $\mathcal{Q}$ in 9.19 is complicated when viewed in terms of Pontrjagin classes,

$$\begin{aligned} q_4 &= p_4 \\ q_8 &= -(p_8 + 2p_4^2) \\ q_{12} &= p_{12} + 18p_4p_8 + 28p_4^3 \\ q_{16} &= -p_{16} - 40p_4p_{12} - 69p_8^2 + 552p_4^2p_8 - 839p_4^4 . \end{aligned}$$

In general $q_{4n} = (-1)^{n-1}p_{4n} + \cdots$ where the dots indicate decomposable terms.

In Chapter 4E we defined an H-equivalence

$$1{+}8\sigma : G/PL[\tfrac12] \to BSO^{\otimes}[\tfrac12]$$

with the property that $ph(1{+}8\sigma) = \hat{j}^*(\mathcal{L})$ where $\hat{j}: G/PL \to BPL$ is the natural map and $\mathcal{L}$ is the PL Hirzebruch class. The results on $F^*(BSO^{\otimes})$ then translate to $F^*(G/PL) \otimes \mathbf{Z}[\tfrac12]$.

THEOREM 9.21. *There is a genus* $\mathfrak{M}$ *for* $F^*(G/PL)\otimes \mathbf{Z}[\tfrac12]$ *such that*

$$F^*(G/PL) \otimes \mathbf{Z}[\tfrac12] = P\{m_4, m_8, \cdots\}$$

and $<\mathfrak{M}, \tilde{\iota}_{4n}> = 8 \cdot (2n-1)!$ *where* $\tilde{\iota}_{4n} \in \pi_{4n}(G/PL)$ *represents the generator with surgery obstruction* 1 (*cf.* 2.25 *and* 4.9).

Proof. The homomorphism

$$(1+8\sigma)_*: \pi_{4r}(G/PL) \to \pi_{4r}(BSO^{\otimes}[\tfrac{1}{2}])$$

takes the generator $\tilde{\iota}_{4r}$ with surgery obstruction 1 into $8/a_n$ times the standard generator ι_{4r} of $\pi_{4r}(BSO^{\otimes}) \otimes Z[\tfrac{1}{2}]$. This is clear from 4.9, 4.22 and the fact that ph evaluated on the standard generator is a_n.

We set $\mathfrak{M} = (1+8\sigma)^*(\mathfrak{Q})$ where $\mathfrak{Q}$ is the genus from 9.19. Then

$$\begin{aligned} <\mathfrak{M}, \tilde{\iota}_{4r}> &= <\mathfrak{Q}, (1+8\sigma)_*(\tilde{\iota}_{4r})> \\ &= 8/a_r <\mathfrak{Q}, \iota_{4r}> \\ &= 8 \cdot (2r-1)!\ . \end{aligned}$$

COROLLARY 9.22. *A sufficient condition that a genus* $\mathfrak{A} \in F^*(G/PL; \bar{Q})$ *be* $Z[\tfrac{1}{2}]$ *integral is that*

$$\pi_r = \frac{1}{8 \cdot (2r-1)!} <\mathfrak{A}, \tilde{\iota}_{4r}>$$

be $Z[\tfrac{1}{2}]$ *integral and satisfy the congruences* $\pi_{rp^i} \equiv \pi_{rp^{i-1}} \pmod{p^i}$ *for every odd prime* p.

Proof. Let h_n be the dual basis to $\mathfrak{M} = (m_4, m_8, \cdots)$ as in 9.6. Then $8 \cdot (2n-1)!\, s_n(h_1, \cdots, h_n)$ is the Hurewicz image of $\tilde{\iota}_{4n}$ and 9.22 follows from 9.11.

C. *The 2-local and integral structure of* $F^*(G/PL)$ *and* $F^*(G/TOP)$

In Chapter 4A we constructed the classes K_{4n} in $F^{4n}(G/PL; Z_{(2)})$ and used them to calculate the 2-local homotopy type of $G/PL[2]$. From 4.8 and 4.9 we obtain the Hopf algebra $F^*(G/PL; Z_{(2)})$. Indeed

9.23
$$F^*(G/PL; Z_{(2)}) = P\{K_4, K_8, \cdots, K_{4n}, \cdots\}$$
$$\psi(K_{4n}) = 1 \otimes K_{4n} + K_{4n} \otimes 1 + 8 \sum_{i=1}^{n-1} K_{4i} \otimes K_{4(n-i)} .$$

From 4.36 we have the natural isomorphism

$$F^*(G/TOP; Z_{(2)}) \xrightarrow{\cong} F^*(G/PL; Z_{(2)}) ,$$

so 9.23 remains valid upon substituting G/TOP for G/PL.

The total class $1 + \Sigma K_{4i}$ is not a genus but $1 + \Sigma\, 8K_i$ is, and the elements

$$k_{4n} = 1/8n\ s_n(8K_4, \cdots, 8K_{4n})$$

are primitive generators of $F^{4n}(G/PL; Z_{(2)})$, where s_n denotes the Newton polynomial (cf. 9.2). Thus, $F^*(G/TOP; Z_{(2)})$ is a primitively generated polynomial algebra, and dually $F_*(G/TOP; Z_{(2)})$ is a divided power algebra,

9.24 $$F_*(G/TOP; Z_{(2)}) = \Gamma\{x_4, x_8, \cdots, x_{4n}, \cdots\}$$

(see below for the definition of Γ).

The element k_{4n} (or K_{4n}) of $F^{4n}(G/TOP; Z_{(2)})$ is 'spherical' in that $\langle k_{4n}, \tilde{\iota}_{4n} \rangle = 1$ for the generator $\tilde{i}_{4n} \in \pi_{4n}(G/TOP)$, cf. 4.37. Thus we can take x_{4n} in 9.24 to be the Hurewicz image of $\tilde{i}_{4n}$. For G/PL we again have

$$F_*(G/PL; Z_{(2)}) = \Gamma\{x_4, x_8, \cdots, x_{4n}, \cdots\}$$

but x_4 is no longer spherical; the generator of $\pi_4(G/PL)$ maps onto $2x_4$.

The divided power algebra Γ is the universal free unitary, graded and commutative algebra over $Z_{(2)}$ on a set of generators $y_1, y_2, \cdots$ subject to the existence of an operator

$$\gamma : \Gamma \to \Gamma$$

satisfying $2\gamma(y) = y^2$ for all $y \in \Gamma$. Γ can be constructed as follows

LEMMA 9.25. (a) $\Gamma\{y_{2i+1}\} = E\{y_{2i+1}\}$ *the exterior algebra on an odd dimensional generator and* $\gamma \equiv 0$.

(b)
$$\Gamma\{y_{2i}\} = P\{y_{2i}, y_{4i}, \cdots, y_{2^j i}, \cdots\}$$

subject to the relations

$$(y_{2^j i})^2 = 2y_{2^{j+1} i} \quad \textit{and} \quad \gamma(y_{2^j i}) = y_{2^{j+1} i}$$

(c)
$$\Gamma\{y_1, \cdots, y_r, y_{r+1}\} = \Gamma\{y_1, \cdots, y_r\} \otimes \Gamma\{y_{r+1}\}.$$

Note that in case (b) $\Gamma\{y_{2i}\}_{2ij}$ is a single copy of $Z_{(2)}$ and $\Gamma\{y_{2i}\}_k = 0$ for other values of k. Moreover, the generator $\gamma_j(y_{2i})$ of $\Gamma\{y_{2i}\}_{2ij}$ can be so chosen that

$$\gamma_j(y_{2i}) \cdot \gamma_s(y_{2i}) = \binom{j+s}{j} \gamma_{j+s}(y_{2i})$$

where $\gamma_0(y_{2i}) = 1$ and $\gamma_1(y_{2i}) = y_{2i}$.

One can give $\Gamma\{y_{2i}\}$ the structure of a Hopf algebra by specifying

$$\psi(\gamma_k(y_{2i})) = \sum_{j=0}^{k} \gamma_j(y_{2i}) \otimes \gamma_{k-j}(y_{2i}).$$

Then $\Gamma\{y_{2i}\}$ is dual to the Hopf algebra $P\{y_{2i}^*\}$ with y_{2i}^* primitive. If we consider $\Gamma\{x_4, x_8, \cdots, x_{4n}, \cdots\}$ as the tensor product of the Hopf algebras $\Gamma\{x_{4i}\}$ then 9.24 above is an isomorphism of Hopf algebras.

We have described $F_*(G/TOP) \otimes Z[½]$ and $F_*(G/TOP) \otimes Z_{(2)}$ and we next want to obtain a description of the integral structure. To this end we need some fairly straightforward algebraic notions which we now develop.

Let A and B be non-trivial unitary subrings of the rationals Q and set $C = A \cap B$. Let M be a free graded A-algebra, N a free graded B-algebra, and let

$$\phi : M \otimes_C N \to M \otimes_A Q$$

be a homomorphism of algebras for which $\phi(x \otimes 1) = x \otimes 1$. To begin we are interested in specifying the kernel K of ϕ. For example, M and N could be the polynomial algebras $M = A[x]$, $N = B[x]$ in a variable x of degree 2 and $\phi(x \otimes 1) = \phi(1 \otimes x) = x$. Then $K(\phi)$ is the ideal generated by $x \otimes 1 - 1 \otimes x$ in $M \otimes_C N$.

In our usual application $A = \mathbf{Z}[1/p_1, 1/p_2, \cdots]$ for some set $S_A = \{p_1, p_2, \cdots\}$ of primes in $\mathbf{Z}$, and B, C have similar representations. In fact $S_A \cap S_B = S_C$. Clearly we have $D = A \otimes_C B$ is again a unitary subring of $\mathbf{Q}$ and $S_D = S_A \cup S_B$.

Let $n_1, \cdots, n_t$ generate N as an algebra over B. Then tensoring with $\mathbf{Q}$ we see that

$$K(\phi \otimes \mathbf{Q}) = I(\cdots (1 \otimes n_t - \phi(1 \otimes n_t) \otimes 1) \cdots)$$

and we have

$$K(\phi) = K(\phi \otimes \mathbf{Q}) \cap M \otimes_C N .$$

DEFINITION 9.26. The equalizer of ϕ is the intersection $K(\phi) \cap (M \otimes_C 1 + 1 \otimes_C N)$ which we write $E(\phi)$.

Thus, in our example above

$$E(\phi) = \bigoplus_i C(x^i \otimes 1 - 1 \otimes x^i) .$$

The following lemma is again clear.

LEMMA 9.27. *$E(\phi)$ is a free C module and (under projection) a subalgebra of $1 \otimes_C N$.*

We find it helpful to think of $E(\phi)$ heuristically as the intersection of M and N.

In Chapter 4 we saw that G/PL is the fiber in

$$G/PL \longrightarrow G/PL[2] \times G/PL[\tfrac{1}{2}] \xrightarrow{8K\times(-ph)} \Pi K(Q, 4i)$$

and similarly for G/TOP. Using the notation above we have

COROLLARY 9.28. $F_*(G/PL) = E[(8K\times(-ph))_*]$ *with a similar result holding for* G/TOP.

Here recall that $F_*(X[2]) = F_*(X)\otimes \mathbf{Z}_{(2)}$ and $F_*(X[\tfrac{1}{2}]) = F_*(X)\otimes \mathbf{Z}[\tfrac{1}{2}]$ and that $F_*(X\times Y) = F_*(X)\otimes F_*(Y)$ so that 9.28 makes sense.

CHAPTER 10

THE TORSION FREE COHOMOLOGY OF BTOP AND BPL

In this and the next chapter we examine the torsion free universal characteristic classes for topological and PL bundles. Each of the two natural maps $\tilde{j}: BO \to BTOP$ and $\hat{j}: G/TOP \to BTOP$ defines a rational equivalence, since their fibers TOP/O and G have finite homotopy groups by results due essentially to Kervaire-Milnor [61], Hirsch-Mazur [53] and Serre [122]. Thus we have two independent calculations of $F_*(BTOP; \mathbf{Q})$, and we will want to refine the considerations to p-local coefficients. It will turn out that $F_*(BTOP; \mathbf{Z}_{(p)})$ is multiplicatively generated by the images of $\tilde{j}_*$ and $\hat{j}_*$, and much of our work in this chapter will be to specify which generators come from BO and which come from G/TOP. For $p = 2$ the two types of generators are different in nature: the generators from BO are polynomial generators but the generators from G/TOP are divided polynomial generators (cf. Chapter 9.C). We shall see that the torsion free cohomology ring $F^*(BTOP)$ is a polynomial ring with one generator in each degree 4n, but the coalgebra structures at 2 and away from 2 are quite different (as one would expect from the above remarks). In the description of $F^*(BTOP)$ as a subring of $H^*(BTOP; \mathbf{Q})$ it is thus convenient to separate the two cases: we treat the embedding $F^*(BTOP; \mathbf{Z}_{(2)}) \subset H^*(BTOP; \mathbf{Q})$ in 10.B below but defer the embedding $F^*(BTOP; \mathbf{Z}[½]) \subset H^*(BTOP; \mathbf{Q})$ to Chapter 11.

A. *The map* $j_*: F_*(BO) \otimes F_*(G/TOP) \to F_*(BTOP)$

The usual maps $\tilde{j}: BO \to BTOP$ and $\hat{j}: G/TOP \to BTOP$ are infinite loop maps by results of Boardman and Vogt, and so too are the maps

$$\pi: BTOP \to BG$$

$$p: BG \to B(G/O)\ .$$

THEOREM 10.1. *The fiber of the composite*

$$\rho: BTOP \to BG \to B(G/O)$$

is BO $\times$ G/TOP *and the inclusion of the fiber is the composite*

$$j : BO \times G/TOP \xrightarrow{\tilde{j} \times \hat{j}} BTOP \times BTOP \xrightarrow{u} BTOP$$

where u *is the Whitney sum map.*

Proof. Consider the diagram

10.2

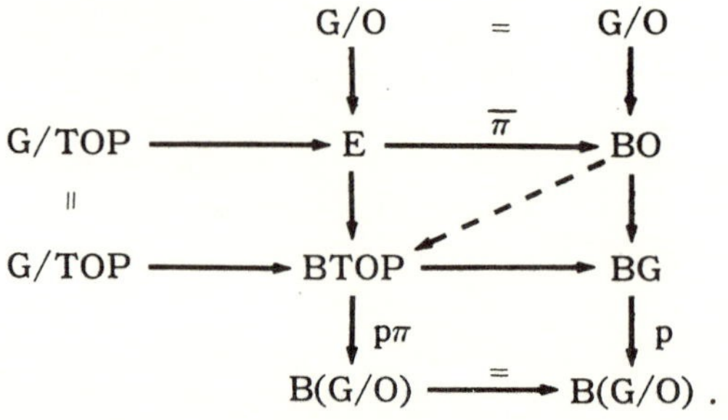

The vertical and horizontal lines are fiberings and $E \to BTOP$ is induced from $BO \to BG$ or from $BTOP \to BG$. Now BO is the fiber of p and the fact that $BO \to BG$ factors through BTOP implies the existence of a one sided homotopy inverse to $\overline{\pi}$, $\theta : BO \to E$. On the other hand the fibering $G/TOP \to E \to BO$ is a fibering of loop spaces so the lifting θ induces $G/TOP \times BO \to E \times E \xrightarrow{u} E$ and it is direct to see that this induces isomorphisms in homotopy. Since all the spaces in question are the homotopy types of CW complexes 10.1 now follows.

COROLLARY 10.3. *The composite*

$$G/O \xrightarrow{\phi} BO \times G/TOP \longrightarrow BTOP$$

is a fibering where ϕ *is the composite*

$$G/O \xrightarrow{\Delta} G/O \times G/O \xrightarrow{r \times \chi s} BO \times G/TOP$$

and r, s *are the usual inclusions while* χ *is the anti-automorphism* $x \to x^{-1}$.

An exactly similar argument works for BPL and we have

COROLLARY 10.4. *There is an infinitely deloopable diagram of fiberings*

$$\begin{array}{ccccccccc}
\longrightarrow & G/O & \xrightarrow{\phi_1} & BO\times G/PL & \xrightarrow{j} & BPL & \longrightarrow & B(G/O) & \longrightarrow \\
 & \Big\downarrow{=} & & \Big\downarrow & & \Big\downarrow & & \Big\downarrow{=} & \\
\longrightarrow & G/O & \xrightarrow{\phi_2} & BO\times G/TOP & \xrightarrow{j} & BTOP & \longrightarrow & B(G/O) & \longrightarrow
\end{array}$$

where the vertical arrows are the usual inclusions.

In particular the maps from BPL and BTOP to B(G/O) are classified by maps

10.5
$$B\phi_1 : B(G/O) \to B^2O\times B(G/PL)$$
$$B\phi_2 : B(G/O) \to B^2O\times B(G/TOP)$$

where for example $B\phi_2$ is the composite

10.6 $$B(G/O) \xrightarrow{\ \Delta\ } B(G/O)\times B(G/O) \xrightarrow{Br\times\chi Bs} B^2O\times B(G/TOP)\ .$$

Our next object in this section is to calculate the image of j_*. The main result of Chapter 13 shows that this image is exactly $F_*(BTOP)$ (cf. Chapter 8.A). With G/TOP replaced by G/PL the same result will also be verified for $F_*(BPL)$.

We have $F_*(BO\times G/TOP) = F_*(BO)\otimes F_*(G/TOP)$ and we note that in each dimension 4n in $F_*(BO)$ or $F_*(G/TOP)$ the set of primitives is a single copy of $\mathbf{Z}$. Let $s_{4n}(G/TOP)$ be a generator in $F_*(G/TOP)$ and $s_{4n}(BO)$ a generator in $F_*(BO)$.

THEOREM 10.7. *Let* K *be the kernel of* j_*, *then* K *has rational generators*

$$\gamma_n = 2^{\alpha(n)-\sigma(n)-1}(2^{2n-1}-1)\,\mathrm{Num}(B_{2n}/4n)\,1\otimes s_{4n}(G/TOP) - 2^{3-\sigma(n)}\,\theta(n)\,s_{4n}(BO)\otimes 1$$

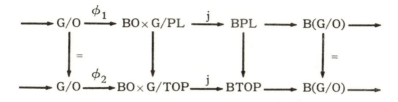

where $\sigma(n) = \max(3, \alpha(n) - 1)$ *) *and* $\theta(n)$ *denotes the odd part of* $\mathrm{Denom}(B_{2n}/4n)$.

Proof. It suffices to show y_n belongs to K since rationally the primitives must generate $F_*(BTOP) \otimes Q$. The Hurewicz image of a generator in $\pi_{4n}(BO)$ is $a_n(2n-1)!\, s_{4n}(BO)$, cf. the proof of 9.18. Since G/TOP is 2-locally a product of Eilenberg-MacLane spaces, the Hurewicz image of the generator of $\pi_{4n}(G/TOP)$ is an odd multiple of $s_{4n}(G/TOP)$, in fact $\mathrm{odd}[(2n-1)!] s_{4n}(G/TOP)$ by 9.21. On the other hand from Chapter 5.E we see that

10.8 $$N_n\, \mathrm{odd}[(2n-1)!] \cdot 1 \otimes s_{4n}(G/TOP) = a_n D_n (2n-1)!\, s_{4n}(BO) \otimes 1$$

where

$$N_n = a_n\, 2^{2n-2}(2^{2n-1}-1)\, \mathrm{Num}(B_{2n}/4n)$$
$$D_n = \mathrm{Denom}(B_{2n}/4n)\,.$$

Now $\nu_2(n!) = n - \alpha(n)$ and from ([2], II, p. 139) $\nu_2(D_n) = \nu_2(n) + 3$, so 10.7 follows from 10.8 when we cancel out the common factor $a_n(2n-1)!\, 2^{\nu_2(n)+\sigma(n)}$.

Having in this way specified K in principle we can calculate the image of j_*. It seems best, however, to do the calculation separately at the various primes.

We begin with the prime 2.

Let $x_{4n} \in F_{4n}(G/TOP)$ be the spherical generator and let $y_{4n} \in F_{4n}(BO)$ be the generator of $F_*(BO)$ given by

$$\langle p_4^n, y_{4n} \rangle = 1$$

$$\langle p_4^{i_1} \cdots p_{4n}^{i_n}, y_{4n} \rangle = 0$$

---

*) Here $\alpha(n)$ is the number of non-zero terms in the dyadic expansion of n.

where $p_4, p_8, \cdots$ are the Pontrjagin classes, cf. Chapter 9.A. Then y_{4n} generates the image of $F_{4n}(BSO_2) \to F_{4n}(BSO)$.

THEOREM 10.9. $\mathrm{Im}(j_*) \otimes \mathbf{Z}_{(2)} = P\{y_{4n} \mid \alpha(n) - 4 < \nu_2(n)\} \otimes$
$\Gamma\{x_{4n} \mid \alpha(n) - 4 \geq \nu_2(n)\}$

where x_{4n}, y_{4n} *are the images in* $F_*(BTOP) \otimes \mathbf{Z}_{(2)}$ *of the generators defined above.*

Proof. In the proof we use the Witt vector basis $\{h_{m,i} \mid m \equiv 1 \pmod 2, i \geq 0\}$ for $F_*(BO) \otimes \mathbf{Z}_{(2)}$ described in Chapter 9.A. The degree of $h_{m,i}$ is $4(2^i m)$ and it is inductively determined by the formulae

10.10
$$s_{2^{i+2}m} = 2^i h_{m,i} + 2^{i-1} h_{m,i-1}^2 + \cdots + h_{m,0}^{2^i}$$

where $s_{2^{i+2}m}$ is the Newton polynomial in the classes $y_4, y_8, \cdots$. We divide the proof into three cases according to the sign of $\alpha(n) - 4$.

Case 1. $(\alpha(n) < 4)$. Then the element γ_n in 10.7 is

$$\gamma_n = 2^{\alpha(n)-4}(2^{2n-1}-1)\,\mathrm{Num}(B_{2n}/4n)(1 \otimes x_{4n}) - \theta_n(s_{4n} \otimes 1)\,.$$

Thus in $F_*(BTOP) \otimes \mathbf{Z}_{(2)}$, x_{4n} is a multiple of s_{4n},

$$x_{4n} \sim 2^{4-\alpha(n)}\,\varepsilon\, s_{4n}\ {}^{*)}\,,$$

and we may suppress x_{4n} from $\mathrm{Im}(j_*)$. Note also that

$$x_{4n}^2 \sim 4 \cdot 2^{6-2\alpha(n)} s_{4n}^2\,,$$

or using the divided power

$$\gamma(x_{4n}) \sim 2 \cdot 2^{6-2\alpha(n)} s_{4n}^2\,,$$

---

*) We use the notation $x \sim y$ here to mean $x = u \cdot y$ for some unit $u \in \mathbf{Z}_2$.

and we can suppress $\gamma(x_{4n})$ as well. This can be iterated so that we can suppress all of

$$\Gamma\{x_{4n} \mid \alpha(n) < 4\} \quad \text{in} \quad \mathrm{Im}(j_*) \otimes \mathbf{Z}_{(2)} .$$

Case 2. $(\alpha(n) = 4)$. Here $s_{4n} = h_{n,0}$ for n odd and $h_{n,0} \sim x_{4n}$.[*)] However, on applying γ, $h^2_{n,0} \sim 2\gamma(x_{4n})$, $h^4_{n,0} \sim 8\gamma^2(x_{4n})$ etc., so we may suppress $P\{h_{n,0}\}$ replacing it by $\Gamma\{x_{4n}\}$.

Case 3. $(\alpha(n) > 4)$. Then $2^{\alpha(n)-4} x_{4n} \sim s_{4n}$ and inductively applying 10.10 gives

(a) If $j \le \alpha(n) - 4$ then there is a $y \in \Gamma\{x_{4n}, \cdots, x_{2^{j+1}n}\}$ so that

$$2^{\alpha(n)-4-j}[x_{2^{j+2}n} + y] \sim h_{n,j}, \quad n \text{ odd} .$$

(b) If $j > \alpha(n) - 4$ then

$$x_{2^{j+2}n} \sim 2^{j-\alpha(n)+4}[h_{n,j} + z]$$

where $z \in \Gamma\{x_{4n}, \cdots, x_{2^{\alpha(n)-2}n}\} \otimes P\{h_{n,\alpha(n)-3}, \cdots, h_{n,j-1}\}$.

It follows that the subalgebra

$$\Gamma\{x_{4n} \mid \alpha(n) - 4 \ge \nu_2(n)\} \otimes P\{h_{n,j} \mid \alpha(n) - 4 < j\}$$

maps onto $\mathrm{Im}(j_*)$. There are no further relations since any such would imply relations on tensoring with $\mathbf{Q}$, but after tensoring with $\mathbf{Q}$ this becomes a polynomial algebra and 10.9 follows.

On localizing at an odd prime p we have

THEOREM 10.11 (Sullivan). *The image of* $F_*(BO;\mathbf{Z}_{(p)}) \otimes F_*(G/TOP;\mathbf{Z}_{(p)})$ *in* $F_*(BTOP;\mathbf{Z}_{(p)})$ *is*

$$P\{y_{2(p-1)m} \mid m > 0\} \otimes P\{x_{4n} \mid n \text{ prime to } p-1\}$$

where y_{4i} *is the polynomial generator from* $F_*(BO) \otimes \mathbf{Z}_{(p)}$ *while* x_{4n} *comes from* $F_*(G/TOP) \otimes \mathbf{Z}_{(p)}$.

Proof. The result follows from the splitting

$$BTOP[p] \simeq BO[p] \times BcokJ_p$$

discussed in Chapter 5.A. Alternately, using the Witt vector basis and facts which follow from the von Staudt theorem

10.12
$$\nu_p(\mathrm{Denom}(B_{2n}/4n)) = 1 + \nu_p(n), \quad 2n \equiv 0 \pmod{p-1}$$
$$\nu_p(\mathrm{Denom}(B_{2n}/4n)) = 0, \quad 2n \not\equiv 0 \pmod{p-1}$$

(see [2], II), the result follows from 10.7.

THEOREM 10.13. *The map* j_* *is onto* $F_*(BTOP)$.

Proof. The space $BcokJ_p$ has finite homotopy groups in each dimension. Hence $\tilde{F}_*(BcokJ_p) = 0$ and the splitting of $BTOP[p]$ implies that j_* is onto at odd primes. At the prime 2 we show in Chapter 13 that in the Bockstein spectral sequence converging to $F_*(BTOP) \otimes \mathbf{Z}/2$ the E^∞-term consists of a polynomial algebra in the stated dimension coming from BO together with divided power algebras in the remaining dimensions coming from G/TOP. This together with 10.9 implies the result at 2 and an easy fitting together argument gives the result over the integers.

The natural map $F_*(BPL) \to F_*(BTOP)$ is an isomorphism (cf. the proof of 8.7), so we also have that

10.14
$$j_* : F_*(BO) \otimes F_*(G/PL) \to F_*(BPL)$$

is surjective.

COROLLARY 10.15. $F^*(BPL) \cong F^*(BTOP)$ *is a polynomial algebra with one generator in each dimension* $4n$.

B. *The embedding of* $F^*(BTOP; Z_{(2)})$ *in* $H^*(BTOP; Q)$

From 10.15 we know that $F^*(BTOP; Z_{(2)})$ is a polynomial algebra with one generator in each dimension $4n$. The Pontrjagin classes $p_{4n} \in H^{4n}(BTOP; Q)$ are not in general $Z_{(2)}$-integral. Instead we exhibit polynomials T_{4n} in the p_{4i} which give polynomial generators for the subring $F^*(BTOP; Z_{(2)})$, but the procedure is involved and the answer relatively unilluminating so we shall be brief. The basic reason for this is that since the dual algebra $F_*(BTOP; Z_{(2)})$ is not polynomial it is not possible to choose the T_{4n} so that the total class $\mathcal{T} = 1 + T_4 + T_8 + \cdots$ is multiplicative.

We begin by calculating the Pontrjagin classes of the index 8 Milnor manifold (cf. 2.16 and Chapter 8.B). Its tangent bundle $\tau(M^{4n})$ is fiber homotopy trivial so its tangent bundle map

$$\tau : M^{4n} \to BPL$$

factors through a map $\bar{\tau}: M^{4n} \to G/PL$. Indeed there is a commutative diagram

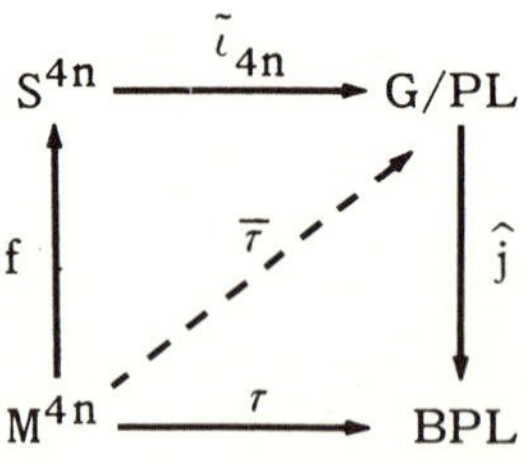

where f has degree 1 and $\tilde{\iota}_{4n}$ represents the generator of $\pi_{4n}(G/PL)$. Thus $\bar{\tau}[M^{4n}] = x_{4n}$, $n > 1$.

LEMMA 10.16. *The total Pontrjagin class of* M^{4n} *is*

$$1 + (2^{4-\alpha(n)}u)\,e$$

where u *is a* $\mathbf{Z}_{(2)}$ *unit and* $<e,[M^{4n}]> = 1$. *(Actually,* $u = \mathrm{odd}[(2n-1)!\ \mathrm{Denom}(B_{2n}/4n)/(2^{2n-1}-1)\,\mathrm{Num}(B_{2n}/4n)]$ *.)*

Proof. The coefficient of p_{4n} in the Hirzebruch class L_{4n} is $(2^{2n+1}(2^{2n-1}-1)/(2n-1)!)\,B_{2n}/4n$ and this is a 2-local integer with 2-adic valuation equal to $2^{a(n)-1}$. Thus we have (cf. Chapter 4.C)

$$
\begin{aligned}
8 = \mathrm{Index}(M^{4n}) &= <\tau^*(\mathcal{L}),[M^{4n}]> \\
&= <\hat{j}^*(\mathcal{L}), x_{4n}> \\
&= <(2^{2n+1}(2^{2n-1}-1)/(2n-1)!)\cdot(B_{2n}/4n)\,p_{4n}, \hat{j}_*(x_{4n})> \\
&= <2^{a(n)-1}\,u\,p_{4n},[M^{4n}]>
\end{aligned}
$$

and 10.16 follows.

COROLLARY 10.17. $<p_{j_1}\cdots p_{j_r}, \gamma^i(x_{4n})>$

$= 0$ *if there is any* s *so* n *does not divide* j_s

$= 2^{2^i(3-a(n))+a(i_1)+\cdots+a(i_r)}$ *if* $j_1 = 4i_1 n, \cdots, j_r = 4i_r n$.

Proof. $\tau_*[(M^{4n})^{2^i}] = 2^{2^i-1}\,\gamma^i(x_{4n})$ so it suffices to calculate

$$<p_{j_1}\cdots p_{j_r},[(M^{4n})^{2^i}]> .$$

The total Pontrjagin class $p((M^{4n})^{2^i})$ is calculated as

$$
\begin{aligned}
p((M^{4n})^{2^i}) &= (1+2^{4-a(n)}ue_1)(1+2^{4-a(n)}ue_2)\cdots(1+2^{4-a(n)}ue_{2^i}) \\
&= 1+\sigma_1+\cdots+\sigma_{2^i}
\end{aligned}
$$

where $\sigma_j \in H^{4nj}((M^{4n})^{2^i},\mathbf{Q})$. Note that up to multiplication by a unit in $\hat{\mathbf{Z}}_2$

$$\sigma_r\sigma_j = \binom{r+j}{j}\sigma_{r+j}$$

for $r+j \leq 2^i$. Hence

$$p_{4i_1 n} \cdots p_{4i_r n} = 2^{2^i(4-\alpha(n))} u^{2^i} \binom{2^i}{i_1, \cdots, i_r} e_1 \otimes \cdots \otimes e_{2^i} .$$

The multinomial coefficient satisfies

$$\binom{2^i}{i_1, \cdots, i_r} = 2^{\alpha(i_1)+\alpha(i_2)+\cdots+\alpha(i_r)-1} \quad \text{(odd)}$$

so that 10.17 follows.

EXAMPLE 10.18. For $n < 15$ all the Pontrjagin numbers on elements in $\Gamma\{x_4, \cdots, x_{56}\}$ evaluate as 2-adic integers. However, for $n = 15$ we have

$$< p_{120}, \gamma(x_{60})> = ½u^2 , \quad < p_{60}^2, \gamma(x_{60})> = u^2 .$$

Likewise for $\gamma^2(x_{60})$ we have 5 numbers, 4 of which are not $Z_{(2)}$ integral.

$$< p_{60}^4, \gamma^2(x_{60})> = 3u^4 , \quad < p_{60}^2 p_{120}, \gamma^2(x_{60})> = \frac{3}{2} u^4 ,$$

$$< p_{180} p_{60}, \gamma^2(x_{60})> = ½u^4 , \quad < p_{120}^2, \gamma^2(x_{60})> = 3/4\, u^4$$

$$< p_{240}, \gamma^2(x_{60})> = \frac{1}{8} u^4 .$$

REMARK 10.19. In view of 10.9 and 10.13 if we wish to find out which $Z_{(2)}$ linear combinations of the Pontrjagin classes are $Z_{(2)}$ integral we need only evaluate them on $\Gamma = \Gamma\{x_{4j} \mid \alpha(j) - 4 \geq \nu_2(j)\}$; they belong to $F^*(BTOP) \otimes Z_{(2)}$ if and only if they evaluate as 2-adic integers on each element in Γ.

The next lemma gives us further information on the structure of the primitives in $F_*(BTOP; Z_{(2)})$.

LEMMA 10.20. *Let* $s_{2^{j+2}n}$ *be the primitive generator in dimension* $2^{j+2}n$ *in* $F_*(BO; Z_{(2)})$, *then*

(a) $\tilde{j}_*(s_{2^{j+2}n}) = 2^r\mu \cdot \hat{j}_*(x_{2^{j+2}n})$ *where* $r = \alpha(n)-4$ *and* μ *is a unit in* $Z_{(2)}$.

(b) *For* $\alpha(n) \geq 4$, $\frac{1}{2}\hat{j}_*(x_{4n}) \in F_*(BTOP; Z_{(2)})$ *if* $\nu_2(n) > \alpha(n) - 4$.

(c) $\hat{j}_*(x_{4n})$ *is a generator if* $\nu_2(n) \leq \alpha(n) - 4$.

Moreover, the result in (b) *is best possible in that* $\frac{1}{4}\hat{j}_*(x_{4n}) \notin F_*(BTOP; Z_{(2)})$.

Proof. (a) and (c) are immediate from 10.16 and 10.9. We prove (b). Neglecting units and suppressing $\tilde{j}_*, \hat{j}_*$ we have in $F_*(BTOP; Z_{(2)})$

$$2^r\, x_{n,i} = 2^i h_{n,i} + \cdots + h_{n,0}^{2^i}$$

where $\{h_{n,i}\}$ is the Witt vector basis for $F_*(BO; Z_{(2)})$ and $x_{n,i} = x_{2^{i+2}n}$. In particular $2^r x_{4n} = h_{n,0}$ and by induction for $i \leq r$ we have

$$2^i h_{n,i} \equiv 2^r\, x_{n,i}\ (\text{mod. } 2^{2r+2-i})$$

(this implies $h_{n,i} \equiv 2^{r-i}\, x_{n,i}\ (\text{mod. } 2^{2r+2-2i})$ and $h_{n,i}^2 \equiv 2^{2(r-i)+1}\gamma(x_{n,i})\ (\text{mod. } 2^{2r+2-2i})$ hence $2^i h_{n,i}^2 \equiv 0\ (\text{mod. } 2^{2r+2-(i+1)})$. From this the induction follows directly).

In particular $h_{n,r} \equiv x_{n,r}\ (\text{mod. } 4)$. Now $2h_{n,r+1} + h_{n,r}^2 \equiv x_{n,r+1}\ (\text{mod. } 4)$ and since $x_{n,r}^2 = 2\gamma(x_{n,r})$ it follows that $x_{n,r+1}$ is divisible by 2. In general, we have

$$2^{i+1}h_{n,r+i+1} + 2^i h_{n,r+i}^2 + \cdots + 2h_{n,r+1}^{2^i} \equiv x_{n,r+i+1}\ (\text{mod. } 4)$$

and (b) follows. Since $h_{n,r+1}$ is a polynomial generator of $F_*(BTOP) \otimes Z_{(2)}$ by 10.9, $\frac{1}{2} x_{n,r+i+1}$ will not be $Z_{(2)}$ integral. This completes the proof.

We are now ready to construct the polynomial generators T_{4n}. We begin with a given odd n. Then associated to n, if $S_{2^{i+2}n}$ represents a primitive generator in $F^*(BO)$, we can construct new elements

$$t_{n,i,j} \in F^*(BO) \otimes \mathbf{Z}_{(2)}$$

according to the formula

10.21 $$S_{2^{i+2}n} = 2^{i-j} t_{n,i,j} + 2^{i-j-1} t_{n,i-1,j}^2 + \cdots + t_{n,j,j}^{2^{i-j}} \qquad (i \geq j).$$

In particular $t_{n,j,j} = S_{2^{j+2}n}$. By induction we see that the $t_{n,i,j}$, $i \geq j$ for a fixed j generate a sub-Hopf algebra of $F^*(BO; \mathbf{Z}_{(2)})$. We identify $H^*(BTOP; Q)$ with $H^*(BO; Q)$ as usual and consider $t_{n,i,j}$ as an element in $H^*(BTOP; Q)$. Then we have

LEMMA 10.22.

(a) $$< t_{n,i,j}, x_{2^{i+2}n} > \; = 2^{j-\alpha(n)+4} \varepsilon$$

(b) $$< t_{n,i,j}, \gamma^t(x_{2^{i-t+2}n}) > \; = 2^{2^t(j-\alpha(n)+3)+1} \varepsilon',$$

if $i-t \geq j$ *and is zero otherwise. Here* $\varepsilon, \varepsilon'$ *are 2-local units.*

Proof. From the Newton formula 9.3 and 10.20 (a)

$$< S_{2^{i+2}n}, x_{2^{i+2}n} > \; = 2^{i-\alpha(n)+4} \varepsilon$$

and (a) follows. Now,

$$\begin{aligned} 0 &= < S_{2^{i+2}n}, (x_{2^{i-t+2}n})^{2^t} > \\ &= 2^{i-j} \left[< t_{n,i,j}, x^{2^t} > + \sum_{\ell=1}^{t} (2^\ell - 1)!\, \varepsilon_\ell < t_{n,i-\ell,j}, x^{2^{t-\ell}} >^{2^\ell} \right] \end{aligned}$$

and since $x^{2^t} = 2^{2^t-1} \gamma^t(x)$, (b) follows directly.

THEOREM 10.23. *Let* n *be an odd number and let* $T_{2^{i+2}n} = T_{n,i} \in H^{2^{i+2}n}(BTOP;Q)$ *be given by*

(i) $T_{n,i} = t_{n,i,0}$ *if* $\alpha(n) - 4 < 0$

(ii) $T_{n,i} = 2^{\alpha(n)-4-i} s_{2^{i+2}n}$ *if* $i \leq \alpha(n) - 4$

(iii) $T_{n,i} = t_{n,i,\alpha(n)-3}$ *if* $i > \alpha(n) - 4$

then these $T_{n,i}$ *are a set of generators for* $F^*(BTOP; \mathbf{Z}_{(2)})$.

Proof. From 10.22a in case (ii)

$$
\begin{aligned}
\langle T_{n,i}, \gamma^r(x_{2^{i-r+2}n}) \rangle &= 0 \quad \text{if} \quad r > 0 \\
&= \varepsilon \quad \text{if} \quad r = 0 .
\end{aligned}
$$

Thus $T_{n,i}$ is integral. Similarly, from 10.22 (b) $T_{n,i}$ is integral in case (iii) and satisfies

$$\langle T_{n,i}, \gamma^t(x_{2^{i-t+2}n}) \rangle = 2\varepsilon' \quad \text{for} \quad i - t \geq \alpha(n) - 3$$

and is zero otherwise.

These calculations imply that the $T_{n,i}$ generate a polynomial sub-Hopf algebra of $F^*(BTOP; \mathbf{Z}_{(2)})$. On the other hand, 10.20 implies the reverse containment and completes the proof.

C. *The structure of* $\Omega_*^{PL}/\mathrm{Tor} \otimes \mathbf{Z}_{(2)}$

The tangent bundle map $\tau_M : M \to BPL$ induces a homomorphism

$$\tau : \Omega_*^{PL}/\mathrm{Tor} \to F_*(BPL) .$$

After tensoring with $\mathbf{Z}_{(2)}$, τ becomes an isomorphism. Indeed, $\tau = \chi_* \circ \Phi \circ h$ where h is the cobordism Hurewicz map from 8.6, and Φ is the Thom isomorphism (8.6).

THEOREM 10.24. *Let* M^{4n} *denote the index* 8 *Milnor manifold of dimension* 4n *and* CP^{2n} *the complex projective space. Then*

$$\Omega_*^{PL}/\mathrm{Tor} \otimes \mathbf{Z}_{(2)} = P\{\{CP^{2n}\} | \nu_2(n) > \alpha(n) - 4\} \otimes \Gamma\{\{M^{4n}\} | \nu_2(n) \leq \alpha(n) - 4\} .$$

Proof. From the paragraph preceding 10.16, $\tau_*\{M^{4n}\} = \hat{j}_*(x_{4n})$ where x_{4n} is the spherical generator of $H_{4n}(G/PL; \mathbf{Z}_{(2)})$ and $\hat{j}: G/PL \to BPL$ the natural map.

Next, the stable tangent bundle of CP^{2n} is $(2n+1)H$ where H is the canonical complex line bundle. The first Chern class of H is a generator of $H^2(CP^{2n})$ so the first Pontrjagin class of its realification $r(H)$ is a generator of $H^4(CP^{2n})$. Since the coefficient of p_4^n in the Newton polynomial $s_n \in H^{4n}(BSO; \mathbf{Z}_{(2)})$ is $1, \langle \tau^*(s_n), [CP^{2n}]\rangle = 1$ and dually $\tau_*\{CP^{2n}\}$ is indecomposable in $H_{4n}(BSO; \mathbf{Z}_{(2)})$. With these remarks 10.24 is direct from 10.9.

The generators in 10.24 are not really explicit, since the divided power operation γ is not a cobordism construction. We now construct a minimal set of generators for $\Omega_*^{PL}/\mathrm{Tor} \otimes \mathbf{Z}_{(2)}$. First, we recall a result from [11], see also [54], p. 13.

LEMMA 10.25. *The* k'th *class* L_k *in the* $\mathcal{L}$*-genus can be written as an integral polynomial with coprime coefficients in the Pontrjagin classes divided by* $\mu(k) = \Pi\, q^{[2k/q-1]}$ *where the product is taken over all primes* q *with* $3 \le q \le 2k+1$.

Let $g(z) = \sqrt{z}/\tanh\sqrt{z} = \Sigma\, b_i z^i$ be the characteristic series for $\mathcal{L}$. It may also be interpreted as $\mathcal{L}(H)$ where $z = e^2$ and H is the canonical line bundle on CP^∞.

DEFINITION 10.26. Let $0 \le k \le m$ and define for a given n the numbers $N_m(n,k)$ by the formula

$$N_m(n,k) = \mu(m)(k^{\text{th}} \text{ coefficient of } g(z)^{2(n+k)+1}).$$

LEMMA 10.27. $N_m(n,k)$ *is an integer.*

Proof. By our remark above the kth coefficient of $g(z)^{2(n+k)+1}$ can be interpreted as $1/\mu(k)\, P_k(p_4, \cdots, p_{4k})$ evaluated on the bundle

$(2(n+k)+1)H$, where the p_{4i} are the Pontrjagin classes, hence integral and P_k has integer coefficients. But $\mu(m)/\mu(k)$ is an integer since $k \leq m$ and 10.27 follows.

We now construct exotic projective spaces CP^{2n+2k} as follows. We start with the degree one normal map

$$f : CP^{2n} \# \mu(m) M_B^{4n} \to CP^{2n} \# S^{4n}$$

and do the construction of Chapter 8.C to get $\widetilde{CP}^{2n+1}$ and induce directly $\widetilde{CP}^{2n+2}$. Then we vary $\widetilde{CP}^{2n+2}$ by $\widetilde{CP}^{2n+2} \# N_m(n,1) \cdot M_B^{4n+4} \to \widetilde{CP}^{2n+2} \# S^{4n+4}$ to get $\widetilde{CP}^{2n+3}$. Proceed as above and vary the resulting $\widetilde{CP}^{2n+4}$ by the obvious normal map with domain $\widetilde{CP}^{2n+4} \# N_m(n,2) M_B^{4n+8}$ etc., until we arrive at $\widetilde{CP}^{2(n+m)}$. Finally, we set

10.28 $$E_{n,m} = \widetilde{CP}^{2(n+m)} \# N_m(n,m) M_B^{4n+4m} .$$

By construction there is a degree one normal map $(\pi_m, \hat{\pi}_m)$ $(\pi_m : E_{n,m} \to CP^{2(n+m)})$, and for each i, π_m is transverse to $CP^{2(n+i)}$ with inverse image $\pi_m^{-1}(CP^{2(n+i)}) = E_{n,i}$. Let

$$f : CP^{2(n+m)} \to G/PL$$

be the map classifying $(\pi_m, \hat{\pi}_m)$. Its restriction to $CP^{2(n+i)}$ classifies $(\pi_i, \hat{\pi}_i)$ which has surgery obstruction $N_m(n,i)$. An inductive application of 4.9 now gives

10.29 $$f^*(K_{4j}) = 0 \quad \text{for} \quad j \neq n$$
$$f^*(K_{4n}) = \mu(m) e^{2n}$$
$$f^*(K_{4j+2}) = 0 .$$

COROLLARY 10.30. *A (multiplicative) basis for* $\Omega_*^{PL}/\mathrm{Tor} \otimes Z_{(2)}$ *is given by the* CP^{2n} *for* $\nu_2(n) > \alpha(n) - 4$, *and the differences* $E_{n,(2^i-1)n} \# - CP^{2^{i+1}n}$ *for* $\nu_2(n) \leq \alpha(n) - 4$, *where* $i \geq 0$ *and* n *is odd.*

Proof. The tangent bundle to $E_{n,m}$ is classified by the composition

$$E_{n,m} \xrightarrow{\pi_m} CP^{2(n+m)} \xrightarrow{\tau \times f} BO \times G/PL \xrightarrow{j} BPL$$

so $\tau(\{E_{n,m}\} - \{CP^{2(n+m)}\}) \in F_*(BPL)$ has the form $j_* f_*[CP^{2(n+m)}] + D$, where D is a decomposable element from $F_*(BO) \otimes F_*(G/PL)$. But 10.29 allows us to calculate f_* on the examples in question. We have

$$< f^*(K^{2^i}_{4n}), [CP^{2^{i+1}n}] > \; = \mu((2^i-1)n)^{2^i},$$

but $\mu(m)$ is odd for all m; hence in $Z_{(2)}$ is a unit and so can be ignored. Thus

$$\tau(\{E_{n,(2^i-1)n}\} - \{CP^{2^{i+1}n}\}) = \gamma^i(x_{4n}) + D$$

where D is decomposable, and 10.30 follows from 10.29 and 10.24.

CHAPTER 11
INTEGRALITY THEOREMS

The first integrality question we consider is to characterize the image of $F^*(\mathrm{BTOP})$ in $H^*(\mathrm{BTOP};\mathbf{Q})$, where as usual we identify the rings $H^*(\mathrm{BTOP};\mathbf{Q})$ and $H^*(\mathrm{BO};\mathbf{Q})$. In Chapter 10.B we treated the 2-local question so it remains to evaluate $F^*(\mathrm{BTOP};\mathbf{Z}[\frac{1}{2}]) \subset H^*(\mathrm{BTOP};\mathbf{Q})$.

Given any topological manifold M^{4n}, let $\tau: M^{4n} \to \mathrm{BTOP}$ be the classifying map of its stable tangent bundle. For any $\alpha \in F^{4n}(\mathrm{BTOP}) \otimes \mathbf{Q}$ we can evaluate α on $\tau_*[M^{4n}]$ to obtain a non-singular pairing

$$\Omega_*^{\mathrm{TOP}}(\mathrm{pt}) \otimes F^*(\mathrm{BTOP};\mathbf{Q}) \to \mathbf{Q}\,.$$

The second integrality question is to find conditions on α so that α actually takes integral values on every manifold.

In part A of this chapter we consider the first question. In part B we consider the second.

A. *The inclusion* $F_*(\mathrm{BTOP};\mathbf{Z}[\frac{1}{2}]) \subset H^*(\mathrm{BTOP};\mathbf{Q})$

We identify $H^*(\mathrm{BTOP};\mathbf{Q})$ with $H^*(\mathrm{BO};\mathbf{Q})$ using $\tilde{j}: \mathrm{BO} \to \mathrm{BTOP}$, and then can give a genus in $H^*(\mathrm{BTOP};\mathbf{Q})$ by specifying its primitive power series as in Chapter 1.E.

From Chapter 10 we have that both $F_*(\mathrm{BTOP};\mathbf{Z}[\frac{1}{2}])$ and $F^*(\mathrm{BTOP};\mathbf{Z}[\frac{1}{2}])$ are polynomial rings, and by 9.12 there is a generating genus for $F^*(\mathrm{BTOP};\mathbf{Z}[\frac{1}{2}])$.

We will show, in fact, in 11.14 that an explicit generating genus is given by $\mathcal{R} = (R_4, R_8, \cdots)$ with primitive series

11.1 $$\mathcal{P}(\mathcal{R}) = \sum (-1)^n (2^{2n-1}-1)\,\mathrm{Num}(B_{2n}/4n)\, z^n$$

where B_{2n} is the Bernoulli number (which, we recall, is defined via the power series expansion

11.2 $$\frac{te^t}{e^t-1} = \sum_{n \geq 0} B_n t^n/n!\ ;$$

in 11.2, $B_0 = 1$, $B_1 = 1/2$, $B_2 = 1/6$ and for $i \geq 1$ $B_{2i+1} = 0$).

Let $j: BO \times G/TOP \to BTOP$ be the map in 10.1. The induced homomorphism

$$j_*: F_*(BO; Z_{(p)}) \otimes F_*(G/TOP; Z_{(p)}) \to F_*(BTOP; Z_{(p)})$$

is surjective for every prime p. Moreover, its image was characterized in 10.9, 10.11. Thus $a \in F^*(BTOP; Q)$ is integral if and only if $\tilde{j}^*(a)$ and $\hat{j}^*(a)$ are both integral.

First we show that $\hat{j}^*(R_{4n})$ is $Z_{(p)}$ integral for all odd primes p.

LEMMA 11.3. *A necessary and sufficient condition that* $\hat{j}^*(\mathcal{R})$ *be* $Z_{(p)}$ *integral is that*

$$\mathrm{Denom}(B_{2np^i}/4np^i) \equiv \mathrm{Denom}(B_{2np^{i-1}}/4np^{i-1})$$

modulo p^i *for all* n *and* i.

Proof. We first calculate $\langle \hat{j}^*(\mathcal{R}), \tilde{\iota}_{4n} \rangle$ where $\tilde{\iota}_{4n} \in \pi_{4n}(G/TOP)$ is the generator with surgery invariant $+1$. Set

11.4
$$D_n = \mathrm{Denom}\,(B_{2n}/4n)$$
$$N_n = 2^{2n-2}(2^{2n-1}-1)\,\mathrm{Num}(B_{2n}/4n)\,a_n$$

and consider the diagram

$$\begin{array}{ccc} G/O & \xrightarrow{\ s\ } & G/TOP \\ \Big\downarrow{\scriptstyle r} & & \Big\downarrow{\scriptstyle \hat{j}} \\ BO & \xrightarrow{\ \tilde{j}\ } & BTOP \end{array}$$

which was evaluated in homotopy in Chapter 5.E. Let $\iota_{4n}(G/O)$ be a generator of $\pi_{4n}(G/O)/\mathrm{Tor}$ and suppose it is so chosen that $r_*(\iota_{4n}(G/O)) = D_n \cdot \iota_{4n}$ where ι_{4n} is the standard generator of $\pi_{4n}(BO)$. Using 4.22 a short calculation gives

$$<8K_{4n}, s_*(\iota_{4n}(G/O))> = D_n \cdot <\mathcal{L}, \iota_{4n}> .$$

Moreover, since the primitive series for the $\mathcal{L}$-genus is

$$P(\mathcal{L}) = \sum (-1)^n \frac{8N_n}{a_n(2n-1)!\,D_n} z^n$$

and since the Pontrjagin class evaluates as $(-1)^{n-1} a_n(2n-1)!$ on ι_{4n} we have

$$s_*(\iota_{4n}(G/O)) = (-1)^{n-1} N_n \cdot \tilde{\iota}_{4n} .$$

Hence, from 11.1 and the commutative square above

$$\begin{aligned} <\hat{j}^*(\mathcal{R}), \tilde{\iota}_{4n}> &= (-1)^{n-1} D_n/N_n <\tilde{j}^*(\mathcal{R}), \iota_{4n}> \\ &= (-1)^{n-1} D_n/2^{2n-2} a_n \cdot <p_{4n}, \iota_{4n}> \\ &= D_n(2n-1)!/2^{2n-2} . \end{aligned}$$

Now, 11.3 follows from 9.22 since the D_n satisfy the required congruences if and only if the $D_n/2^{2n+1}$ do.

LEMMA 11.5. *The congruence of* 11.3 *is valid for all* n, i.

Proof. The denominator D_n of $B_{2n}/4n$ was given in 10.12. If $2n \equiv 0 \pmod{(p-1)}$ the congruence holds trivially. If $2n \not\equiv 0 \pmod{(p-1)}$, suppose $\nu_q(D_{np^i}) \neq 0$ while $\nu_q(D_{np^{i-1}}) = 0$, then $q-1$ divides $2np^i$ but not $2p^{i-1}n$. Hence p^i divides $q-1$ and $q \equiv 1 \pmod{p^i}$, which completes the proof.

LEMMA 11.6. *A sufficient condition that* $\tilde{j}^*(\mathcal{R})$ *be* $\mathbf{Z}_{(p)}$ *integral is that*

$$\mathrm{Num}(B_{2np^i}/4np^i) \equiv \mathrm{Num}(B_{2np^{i-1}}/4np^{i-1})$$

modulo p^i *for all* n *and* i.

Proof. Note first that $2np^i \equiv 2np^{i-1} \pmod{p^i}$, so if the congruences above are satisfied, then the numbers $(2^{2n-1}-1)\,\mathrm{Num}(B_{2n}/4n)$ satisfy the congruences of 9.11, and $\tilde{j}^*(\mathcal{R})$ is $\mathbf{Z}_{(p)}$ integral.

The Kummer congruences of 9.20 together with 11.5 imply the congruences of 11.6 in the case $2n \not\equiv 0 \pmod{(p-1)}$. Before we give the number-theoretic proof that the congruences are also satisfied when $2n \equiv 0 \pmod{(p-1)}$ we digress and prove

LEMMA 11.7. *Suppose* $A_{4n} \in F^*(BTOP;\, Z[\tfrac{1}{2}])$ *is an element with* $A_{4n} = \mu_n \cdot p_{4n} + \cdots$. *Then* A_{4n} *is a polynomial generator if and only if* μ_n *is divisible by precisely* $(2^{2n-1}-1)\,\mathrm{Num}(B_{2n}/4n)$ *in* $Z[\tfrac{1}{2}]$.

Proof. Let k be a positive number which generates $(Z/p^2)^{\times}$. The composition

$$G/TOP[\tfrac{1}{2}] \xrightarrow{\hat{j}} BTOP[\tfrac{1}{2}] \xrightarrow{\rho_L^k} BSO^{\otimes}[\tfrac{1}{2}, 1/k]$$

induces multiplication by $(k^{2n}-1)$ on the homotopy group in dimension 4n (cf. 5.7). Now,

$$(\rho_L^k)^* : F^*(BSO^{\otimes};\, \mathbf{Z}_{(p)}) \to F^*(BTOP;\, \mathbf{Z}_{(p)})$$

is an isomorphism, and neglecting $\mathbf{Z}_{(p)}$ units we have

11.8 $$\langle A_{4n}, \hat{j}_*(\tilde{\iota}_{4n})\rangle = (k^{2n}-1)(2n-1)!$$

if A_{4n} is a generator. On the other hand (as in Chapter 5.E)

$$8 = \langle \mathcal{L}, \hat{j}_*(\tilde{\iota}_{4n}) \rangle$$

$$= \frac{2^{2n+1}(2^{2n-1}-1)}{(2n-1)!}(B_{2n}/4n) \cdot \mu_n^{-1} \langle A_{4n}, \hat{j}_*(\tilde{\iota}_{4n}) \rangle .$$

But $k^{2n} - 1/\mathrm{Denom}(B_{2n}/4n)$ is integral and not divisible by p (see [2] II, p. 139), so substituting in 11.8 we see that μ_n is precisely divisible by $(2^{2n-1}-1)\,\mathrm{Num}(B_{2n}/4n)$ in $Z_{(p)}$. This is true for all odd primes, and so gives the result in $Z[\frac{1}{2}]$. The reverse implication is proved in a similar fashion.

Our original proof of the validity of the congruence in 11.6 was based on topological considerations. We thank Larry Washington for showing us the elementary proof we give below.

LEMMA 11.9. *The congruence of* 11.6 *is valid for all* n, i *with* $p-1$ *dividing* $2n$.

Proof. The Bernoulli polynomials are defined via the power series

11.10 $$\frac{te^{(1+x)t}}{e^t - 1} = \sum_{n \geq 0} B_n(x)\, t^n/n! .$$

In particular, on expanding out we have

11.11 $$B_n(x) = \sum_{j=0}^{n} \binom{n}{j} B_j x^{n-j}$$

where B_j is the jth Bernoulli number.

LEMMA 11.12.

$$pB_n = (-p)^n \sum_{a=1}^{p} B_n(-a/p), \qquad n > 1 .$$

Proof. Consider the series

$$\sum_{n\geq 0} t^n/n! \left\{ p^{n-1} \sum_{a=1}^{p} B_n(-a/p) \right\} = \frac{1}{p} \sum_{a=1}^{p} \sum_{n\geq 0} (tp)^n/n!\, B_n(-a/p)$$

$$= \frac{1}{p} \sum_{a=1}^{p} \frac{pt}{e^{pt}-1}\, e^{(1-a/p)pt}$$

$$= \frac{te^{pt}}{e^{pt}-1} \left(\sum_{a=1}^{p} e^{-at} \right)$$

$$= \frac{te^{pt}}{e^{pt}-1} \left(\frac{e^{-pt}-1}{e^{-t}-1} \right) e^{-t}$$

$$= \frac{-te^{-t}}{e^{-t}-1} = \sum_{n\geq 0} (-1)^n B_n t^n/n!$$

and 11.12 follows.

Expanding 11.12 by using 11.11 we have

11.13 $$pB_n = (-1)^n \sum_{a=1}^{p} \sum_{j=0}^{n} \binom{n}{j} pB_j(-a)^{n-j} p^{j-1} .$$

Note from the proof of 11.5 that $pB_n \in Z_{(p)}$. Now, from 11.13 we have, since $p \geq 3$

$$pB_{np^i} \equiv (-1)^n \sum_{a=1}^{p} (-a)^{np^i} + (-1)^{n/2} \left(\sum_{a=1}^{p} p^i n(-a)^{p^i n-1} \right) \cdot p \pmod{p^{i+1}}$$

since $\nu_p\binom{np^i}{j} \geq i - \nu_p(j)$ and $B_0 = 1$, $B_1 = \frac{1}{2}$. Hence

$$pB_{np^i} \equiv (-1)^n \sum_{a=1}^{p} (-a)^{np^i} \pmod{p^{i+1}} .$$

On the other hand, since $p-1$ divides n, $a^{np^i} \equiv 1 \pmod{p^{i+1}}$ and we have

$$pB_{np^i} \equiv p-1 \pmod{p^{i+1}} .$$

In particular, $pB_{np^i} \equiv pB_{np^{i+1}} \pmod{p^{i+1}}$ and 11.9 follows since

$$\frac{p\mathrm{Denom}(B_{np^i}/2np^i)}{\mathrm{Denom}(B_{np^{i+1}}/2np^{i+1})} \equiv 1 \pmod{p^{i+1}} .$$

THEOREM 11.14. $F^*(BTOP; Z[\frac{1}{2}]) = P\{R_4, R_8, \cdots, R_{4n}, \cdots\}$ *where* $\mathcal{R}$ *is the genus from* 11.1.

Proof. We have seen above that $\mathcal{R}$ is $Z_{(p)}$ integral for every odd prime hence $Z[\frac{1}{2}]$ integral and 11.14 follows from 11.7.

The genus $\mathcal{R}$ is not Z integral except on low dimensional classes. If one wants a genuinely integral genus which generates $F^*(BTOP) \otimes Z[\frac{1}{2}]$ one may for example take

$$1 + 2R_4 + 4R_8 + \cdots + 2^n R_{4n} + \cdots .$$

Also, note that 11.14 remain true with BTOP replaced by BPL.

We end this section by pointing out the conditions under which a genus $\mathfrak{A}$ is $Z_{(2)}$ integral on all PL manifolds.

LEMMA 11.15. $\mathfrak{A}$ *is* $Z_{(2)}$ *integral on all* PL *manifolds if and only if for the primitive series* $P(\mathfrak{A}) = \Sigma(-1)^i \pi_i z^i$ *the conditions* i) *and* ii) *in* 9.11 *are satisfied and for* i *such that* $\alpha(i)-4 \geq \nu_2(i)$ *then* $2^{\alpha(i)-3}$ *divides* π_i.

Proof. Since

$$\tau : \Omega_*^{PL}(pt) \otimes Z_{(2)} \to F_*(BPL) \otimes Z_{(2)}$$

is onto, 11.15 follows from 9.11, 10.9 and 10.20(a) when we note that if $<\mathfrak{a}, x_{4i}>$ is not divisible by 2 then $<\mathfrak{a}, \gamma(x_{4i})> = \frac{1}{2}<\mathfrak{a}, x_{4i}^2>$ does not belong to $Z_{(2)}$. On the other hand, if $<\mathfrak{a}, x_{4i}> \in 2Z_{(2)}$ then $<\mathfrak{a}, \gamma(x_{4i})> \in 2Z_{(2)}$ as well so we can iterate.

B. *Piecewise linear Hattori-Stong theorems*

We now turn to the second integrality question. To begin we recall the smooth Hattori-Stong theorem. The stable tangent bundle $\tau_M : M \to BSO$ induces a monomorphism

$$\tau : \Omega_*^{SO}/\mathrm{Tor} \to H_*(BSO; Q)$$

defining a sublattice $B = \mathrm{Im}\,\tau \subset H_*(BSO; Q)$ completely described by its dual lattice B^*. In describing B or B^* it is convenient to separate the two cases: at 2 and away from 2. That is, we will give a description of B^* as

11.16 $$B^* = B[\tfrac{1}{2}]^* \cap B[2]^*$$

where $B[\frac{1}{2}]^*$, $B[2]^*$ are the sets of homomorphisms $f: H_*(BSO) \to Q$ with $f|B \in Z[\frac{1}{2}]$ or $f|B \in Z_{(2)}$, and B^* is the set of homomorphisms which take integral values on B.

Recall the KO orientation of $MSO[\frac{1}{2}]$ from 4.14,

$$\Delta : MSO[\tfrac{1}{2}] \to BSO[\tfrac{1}{2}]$$

with $\mathrm{ph}\Delta = \mathcal{L}^{-1} \cdot U$. To each $x \in KO(BSO) \otimes Z[\frac{1}{2}]$ and each class $\{M^{4n}\} \in \Omega_*^{SO}/\mathrm{Tor}$ there is associated a characteristic number $x\{M\} \in Z[\frac{1}{2}]$ as follows: Let

$$\begin{array}{ccccc} f_M : S^{4n+4k} & \longrightarrow & M(\nu_M) & \longrightarrow & MSO_{4k} \\ & & \cup & & \cup \\ & & M^{4n} & \xrightarrow{\quad \nu \quad} & BSO_{4k} \end{array}$$

be a classifying diagram for the cobordism class of M^{4n} (cf. Chapter 1.C).

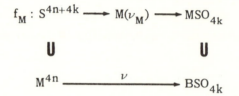

Then

11.17 $$f_M^*(x \cdot \Delta) = x\{M\} \cdot a^{n+k}$$

where $a \in \widetilde{KO}(S^4) \otimes Z[\frac{1}{2}]$ is the generator with $ph_4(a) \in H^4(S^4; Z)$ the standard generator. We can take the Pontrjagin character on 11.17 and use the fact that $ph\Delta = \mathcal{L}^{-1} \cdot U$ to obtain the equivalent cohomological expression

11.18 $$x\{M\} = \langle \mathcal{L}(M) \cdot \nu^*(ph(x)), [M^{4n}] \rangle = \langle \mathcal{L} \cdot ph(\chi^*(x)), \tau_*[M^{4n}] \rangle$$

where $\chi : BSO^{\oplus} \to BSO^{\oplus}$ is the H-space inverse. The K-theoretic definition of $x\{M\}$ shows that $x\{M\} \in Z[\frac{1}{2}]$ and thus that $ph(KO(BSO) \otimes Z[\frac{1}{2}]) \cdot \mathcal{L}$ is contained in $B[\frac{1}{2}]^*$. The opposite containment is due to Hattori and Stong (see e.g. [133], p. 207).

THEOREM 11.19 (Hattori-Stong). $B[\frac{1}{2}]^* = ph(KO(BSO; Z[\frac{1}{2}])) \cdot \mathcal{L}$.

To get further information on the subring $B[\frac{1}{2}]^*$ of $H^*(BSO; Q)$ we must compute $ph\, KO(BSO; Z[\frac{1}{2}])$. To this end consider

$$\gamma_t : KO(X) \to KO(X)[[t]]$$

and define

$$\pi_s : KO(X) \to KO(X)[[s]]$$

by $\pi_s(\xi) = \gamma_t(\xi - \dim \xi)$ where $s = t - t^2$. The coefficients in $\pi_s(\xi)$ are the KO-theoretic Pontrjagin classes of ξ, $\pi_s(\xi) = \Sigma\, \pi^k(\xi)\, s^k$, and

$$KO(BSO; Z[\tfrac{1}{2}]) = Z[\tfrac{1}{2}]\,[[\pi^1, \pi^2, \cdots]]$$

where $\pi^i = \pi^i$ (universal class) [5], [13]. Hence, to compute the subring $ph\, KO(BSO; Z[\frac{1}{2}]) \subset H^*(BSO; Q)$ it suffices to calculate $ph\, \pi^i(\xi)$ in terms of the Pontrjagin classes of ξ.

If $\xi = L_1 + \cdots + L_u$ is a sum of complex line bundles then

$$c \circ \pi_s \circ r(L_1 + \cdots + L_u) = \gamma_t(\Sigma(L_i + \bar{L}_i - 2)) = \Pi(1 + t(L_i - 1))(1 + t(\bar{L}_i - 1)) = $$
$$= \Pi(1 + s(L_i + \bar{L}_i - 2)$$

where c, r denote complexification and realification respectively. So

$$\text{ph}\, \pi_s(L_1 + \cdots + L_u) = \Pi(1 + s(e^{x_i} + e^{-x_i} - 2)) .$$

Thus if we formally write $p(\xi) = \Pi(1 + x_i^2)$ then the Pontrjagin character of $\pi^k(\xi)$ is the k'th elementary symmetric function in the $e^{x_i} + e^{-x_i} - 2$,

$$\text{ph}(\pi^k(\xi)) = \sigma_k(e^{x_1} + e^{-x_1} - 2, \cdots, e^{x_k} + e^{-x_k} - 2) .$$

and ph KO(BSO; $Z[½]$) is the power series ring over $Z[½]$ generated by these elements.

The $Z_{(2)}$ dual lattice of B is much simpler to describe: MSO[2] is a wedge of suspensions of the Eilenberg-MacLane spectra $K(Z_{(2)})$ and $K(Z/2)$ with a single copy of $K(Z_{(2)})$ for each additive generator of $F^*(BSO)$. From 1.18 (a) we get

THEOREM 11.20. $B[2]^* = Z_{(2)}[p_4, p_8, \cdots]$.

We make 11.16 explicit with

COROLLARY 11.21 (Hattori-Stong). *The sublattice* $B = \Omega_*^{SO}/\text{Tor}$ *in* $H_*(BSO; Q)$ *consists of the classes* a *such that*

(i) $\langle g, a \rangle \in Z[½]$ *for* $g \in B[½]^*$

and

(ii) $\langle g, a \rangle \in Z_{(2)}$ *for* $g \in B[2]^*$

where $B[½]^*$ *and* $B[2]^*$ *are described in* 11.19 *and* 11.20.

There are results analogous to 11.19 and 11.20 for the topological and PL cobordism rings.*) Again the stable tangent bundle induces a monomorphism

$$\tau : \Omega_*^{PL}/\mathrm{Tor} \to H_*(BPL; Q)$$

whose image C is a sublattice of $H_*(BPL; Q)$. At each odd prime p we have the splitting from 5.20

$$MSPL[p] \simeq MSO[p] \wedge M\,\mathrm{cok}\, J_p .$$

Let $\gamma_p : BSO[p] \to BSPL[p]$ be the exotic map defined in 5.11. The associated map of Thom spectra $\hat{\gamma}_p : MSO[p] \to MSPL[p]$ is the inclusion in the splitting of $MSPL[p]$. We now restate and prove Theorem 5.21,

THEOREM 11.22. $\hat{\gamma}_{p*} : \pi_*^S(MSO[p])/\mathrm{Tor} \to \pi_*^S(MSPL[p])/\mathrm{Tor}$ *is an isomorphism.*

Proof. We identify $\pi_*^S(MSO[p])$, $\pi_*^S(MSPL[p])$ with $\Omega_*^{SO} \otimes Z_{(p)}$ and $\Omega_*^{PL} \otimes Z_{(p)}$ and consider the diagram

$$\begin{array}{ccc} \Omega_*^{SO}/\mathrm{Tor} \otimes Z_{(p)} & \xrightarrow{(\hat{\gamma}_p)_*} & \Omega_*^{PL}/\mathrm{Tor} \otimes Z_{(p)} \\ \downarrow \tau & & \downarrow \tau \\ H_*(BSO; Q) & \xrightarrow{(\gamma_p)_*} & H_*(BSPL; Q) \end{array}$$

where both vertical maps are monomorphisms with images $B \otimes Z_{(p)}$ and $C \otimes Z_{(p)}$, respectively. Let $\Delta_{PL} \in \widetilde{KO}(MSPL[p])$ be the Thom class from 5.2. Then $\hat{\gamma}_p^*(\Delta_{PL})$ is a KO-theoretic Thom class for $MSO[p]$ and hence $\hat{\gamma}_p^*(\Delta_{PL}) = \gamma^{-1} \cdot \Delta_{SO}$, where Δ_{SO} was characterized in 4.14, and γ is a unit of $KO(BSO[p])$. Taking Pontrjagin characters we get

*) The two cases are identical as the natural map $\Omega_*^{PL}/\mathrm{Tor} \to \Omega_*^{TOP}/\mathrm{Tor}$ is an isomorphism.

11.23 $$\gamma_p^*(\mathcal{L}) = \mathrm{ph}(\gamma)\cdot\mathcal{L}$$

where $\mathcal{L}$ is the Hirzebruch class in either $H^*(BSO; \mathbf{Q})$ or $H^*(BSPL; \mathbf{Q})$.

Consider an element $\{M\} \in \Omega_*^{PL}/\mathrm{Tor}\otimes \mathbf{Z}_{(p)}$. The argument used in 11.18 above gives

$$<\mathcal{L}\cdot \mathrm{ph}(\xi), \tau\{M\}> \in \mathbf{Z}_{(p)}$$

for all $\xi \in KO(BSPL[p])$. Now, $\gamma_p : BSO[p] \to BSPL[p]$ induces isomorphisms in both ordinary rational cohomology and in KO-cohomology by 5.12, 5.18, 5.24. Let $b \in H_*(BSO; \mathbf{Q})$ be the unique element with $(\gamma_p)_*(b) = \tau\{M\}$. We have $\gamma_p^*(\mathcal{L}\cdot \mathrm{ph}(\xi)) = \mathcal{L}\cdot \mathrm{ph}(\gamma\cdot(\gamma_p)^*(\xi))$ and hence

$$<\mathcal{L}\cdot \mathrm{ph}(\eta), b> \in \mathbf{Z}_{(p)}$$

for all $\eta \in KO(MSO[p])$. But then 11.19 shows that $b \in B\otimes \mathbf{Z}_{(p)}$ so $b = \tau\{N\}$, and $\{M\} = (\hat{\gamma}_p)_*\{N\}$. This proves that $(\hat{\gamma}_p)_*$ is surjective, and since the injectivity is obvious, proves the theorem.

As a direct consequence of 11.19 we have

COROLLARY 11.24. *The lattice* $C[\frac{1}{2}] = \mathrm{Im}\{\Omega_*^{PL}/\mathrm{Tor}\otimes \mathbf{Z}[\frac{1}{2}] \to H_*(BSPL{:}\,\mathbf{Q})\}$ *is characterized by*

$$C[\tfrac{1}{2}]^* = \mathrm{ph}(KO(BSPL; \mathbf{Z}[\tfrac{1}{2}]))\cdot\mathcal{L}\ .$$

At the prime 2 we have in analogy with 11.20 (see also the paragraph preceding Lemma 8.6).

THEOREM 11.25. $C_{(2)}^* = \mathbf{Z}_{(2)}[T_4, T_8, \cdots]$ *where the* T_{4i} *are the generators of* $F^*(BSPL)\otimes \mathbf{Z}_{(2)}$ *given in* 10.23.

COROLLARY 11.26 (PL Hattori-Stong Theorem). *The lattice* $C = \tau(\Omega_*^{PL}/\mathrm{Tor}) \subset H_*(BSPL; \mathbf{Q})$ *has dual lattice* $C^* = C[\frac{1}{2}]^* \cap C_{(2)}^*$, *where* $C[\frac{1}{2}]^*$ *and* $C_{(2)}^*$ *are described in* 11.24, 11.25.

REMARK 11.27. In analogy with the smooth case (see the paragraphs following 11.9) one would like to have a more explicit calculation of $G[1/2]^*$. In particular one would like to know ph $KO(BSPL; \mathbb{Z}[1/2]) \subset H^*(BSPL; \mathbb{Q})$ in terms of the genus $\mathcal{R}$ introduced in 11.1. The most satisfactory answer would perhaps be that ph $KO(BSPL; \mathbb{Z}[1/2])$ arises from ph $KO(BSO; \mathbb{Z}[1/2])$ by substituting R_{4i} for p_{4i}. This would happen if there is an element $\xi \in KO(BSPL; \mathbb{Z}[1/2])$ such that

$$ph(\xi) = \sum \frac{2}{(2n)!} s_n(R_4, \cdots, R_{4n}) .$$

Using the inclusion BSO → BSPL one wonders if

$$\sum_{n \geq 1} (2^{2n-1}-1) \mathrm{Num}(B_{2n}/4n) \cdot \frac{2}{(2n)!} s_n(p_4, \cdots, p_{4n}) \in \mathrm{ph}\ KO(BSO; \mathbb{Z}[1/2])$$

or equivalently if one can show

11.28 $$\sum_{n \geq 1} (2^{2n-1}-1) \mathrm{Num}(B_{2n}/4n) \frac{2}{(2n)!} x^{2n} \in \mathrm{ph}\ \tilde{K}(CP^\infty; \mathbb{Z}[1/2]) .$$

This is a purely number theoretic question as

$$K(CP^\infty, \mathbb{Z}[1/2]) = \mathbb{Z}[1/2][[\xi]] \text{ and}$$

$$\mathrm{ph}\ f(\xi) = f(e^x + e^{-x} - 2) .$$

In a similar fashion one can use the inclusion G/PL → BSPL and the equivalence (away from 2) between G/PL and BO to get a second condition, analogous to 11.28, but involving the denominator of $B_{2n}/4n$ rather than the numerator. We leave the details to the reader.

C. *Milnor's criteria for* PL *manifolds*

Recall that a smooth manifold M^{4n} defines an indecomposable cobordism class $\{M^{4n}\}$ in the ring $\Omega_*^{SO}/\mathrm{Tor} \otimes \mathbb{Z}_{(p)}$ if and only if the following criteria (due to Milnor) are satisfied:

11.29
$$\langle s_n, [M^{4n}] \rangle \in \mathbb{Z}_{(p)}^\times \quad \text{if} \quad 2n+1 \neq p^a$$
$$\langle s_n, [M^{4n}] \rangle \in p\mathbb{Z}_{(p)}^\times \quad \text{if} \quad 2n+1 = p^a$$

where s_n is the Newton polynomial in the Pontrjagin classes and $< , >$ denotes the usual pairing

$$F^*(BSO; Z_{(p)}) \otimes \Omega_*/\mathrm{Tor} \to Z_{(p)} .$$

(see e.g. [133]).

There is an analogous statement for PL (or topological) manifolds. Let $s_n(\mathcal{R}) \in F^{4n}(BSPL; Z[½])$ denote the Newton polynomial in the classes $R_4, R_8, \cdots$.

THEOREM 11.30. *Let* p *be an odd prime. A necessary and sufficient condition that a* PL *manifold* M^{4n} *be indecomposable in* $\Omega_*^{PL}/\mathrm{Tor} \otimes Z_{(p)}$ *is that* $< s_n(\mathcal{R}), [M^{4n}] > \in Z_{(p)}^{\times}$ *if* $2n+1 \neq p^a$ *and* $< s_n(\mathcal{R}), [M^{4n}] > \in pZ_{(p)}^{\times}$ *if* $2n+1 = p^a$.

Proof. If $\tilde{j}: BO \to BPL$ and $\hat{j}: G/PL \to BPL$ are the natural maps then

(a) $\tilde{j}^*(s_n(\mathcal{R})) = (2^{2n-1}-1)\, \mathrm{Num}(B_{2n}/4n) \cdot s_n$

(b) $\hat{j}^*(s_n(\mathcal{R})) = 1/2^{2n+1}\, \mathrm{Denom}(B_{2n}/4n) \cdot s_n(\mathfrak{M})$

where $\mathfrak{M}$ is the genus for $F^*(G/PL)$ from 9.21 (cf. p. 211). The coefficient of s_n in (a) is a unit in $Z_{(p)}$ when $2n \equiv 0(p-1)$ and the coefficient in (b) of $s_n(\mathfrak{M})$ is a unit when $2n \not\equiv 0(p-1)$, (cf. 10.12). The proof is now direct from 11.22.

REMARK 11.31. Note as a consequence of 11.30 that in dimensions congruent to zero modulo $(p-1)$ a smooth manifold W^{2n} which is indecomposable in $\Omega_* \otimes Z_{(p)}$ is also indecomposable in $\Omega_*^{PL} \otimes Z_{(p)}$. The remaining generators are less explicit: Let $CP^{2n} \xrightarrow{f} G/PL[½]$ be a map so that $\sigma \circ f: CP^{2n} \to BSO^{\otimes}[½]$ classifies the canonical (complex) line bundle. For a sufficiently large power of 2, $2^e \cdot f$ becomes integral (i.e. maps into G/PL) and $\eta\{CP^{2n}, 2^e \cdot f\} - \{CP^{2n}\}$ is indecomposable in $\Omega_{4n}^{PL} \otimes Z_{(p)}$ when $2n \not\equiv 0(p-1)$.

CHAPTER 12*)

THE SMOOTH SURGERY CLASSES AND $H_*(BTOP; Z/2)$

In this chapter we use the fibrations

12.1
$$BO \times G/TOP \xrightarrow{j} BTOP \longrightarrow B(G/O)$$
$$BTOP \to B(G/O) \to B^2O \times B(G/TOP)$$

of Chapter 10.A and their Leray-Serre spectral sequences to obtain a description of $H_*(BTOP; Z/2)$. Effective calculations in the spectral sequences of 12.1 require information on the mapping $Bs: B(G/O) \to B(G/TOP)$. Specifically, the calculation of $H_*(BTOP; Z/2)$ requires knowledge of the induced map $(Bs)_*$ in mod 2 homology. However, for the calculation of the mod 2 Bockstein spectral sequence of BTOP in the next chapter we will need $Z_{(2)}$ integral information as well. In part A below we review the results from [35] and [78] which completely characterize the map $Bs: B(G/O) \to B(G/TOP)[2]$.

In part B we obtain $H_*(BTOP; Z/2)$. This gives a (non-geometric) description of the unoriented topological cobordism ring in dimensions $\neq 4$ since

12.2
$$\mathfrak{N}_*^{TOP} = \mathfrak{N}_*^{Diff} \otimes H_*(B(TOP/O); Z/2), \quad * \neq 4$$

and $H_*(B(TOP/O); Z/2) = H_*(BTOP; Z/2)//H_*(BO; Z/2)$. (If there exists an almost parallelizable topological 4-manifold of index 8, then 12.2 also holds in dimension 4.)

A. *The map* $B(r \times s): B(G/O) \to B^2O \times B(G/TOP)$.

To begin we recall the structure of $H_*(G/O; Z/2)$ and $H_*(B(G/O); Z/2)$.

*) This chapter and the next are highly technical in nature, and the reader is advised to first glance through them to see if there is a pressing need to learn the techniques outlined here.

THEOREM 12.3. $H_*(G/O; Z/2) = P\{e_2, e_4, \cdots, e_{2i}, \cdots\} \otimes P\{x_{(i_1,\cdots,i_r)} \mid 0 < i_1 \leq i_2 \leq \cdots \leq i_r,\ r \geq 2\}$, *where* $\deg(e_{2i}) = 2i$ and $\deg x_{(i_1,\cdots,i_r)} = i_1 + 2i_2 + \cdots + 2^{r-1} i_r$.

Indeed, in 12.3 x_I is the image in $H_*(G/O; Z/2)$ of the element $\hat{Q}_{i_1} \cdots \hat{Q}_{i_{r-2}}(Q_{i_{r-1}} Q_{i_r}[1] * [-3])$ from 6.25 under the natural map $SG \to G/O$ and e_{2i} is the image of $Q_0 Q_i[1] * [-3]$. Theorem 12.3 follows easily from 6.25 using the Leray-Serre spectral sequence of the fibering $SO \to SG \to G/O$.

The Leray-Serre spectral sequence passing from G/O to $B(G/O)$ is totally transgressive and after taking account of the Dyer-Lashof operation $\hat{Q}_1$ in $H_*(G/O; Z/2)$,

$$\hat{Q}_1(x_{(i_1,\cdots,i_r)}) = x_{(1,i_1,\cdots,i_r)}$$

$$\hat{Q}_1(e_{2i}) = 0$$

we have

THEOREM 12.4. $H_*(B(G/O); Z/2) = E\{f_3, f_5, \cdots, f_{2i+1}, \cdots\} \otimes P\{y_{i,j} \mid 1 \leq i \leq j\} \otimes P\{y_{(i_1,\cdots,i_r)} \mid r > 2,\ 1 < i_1 \leq i_2 \leq \cdots \leq i_r\}$, *where* $\deg(y_I) = 1 + i_1 + 2i_2 + \cdots + 2^{r-1} i_r$ *and each* y_I *is the suspension of the corresponding* x_I *in 12.3 as are the* f_{2i+1} *of* e_{2i}.

(Alternately, the structure of $H_*(B(G/O); Z/2)$ follows from 6.26 upon using the Leray-Serre spectral sequence of $BO \to BG \to B(G/O)$.)

The suspension

$$\sigma_* : H_*(G/O; Z/2) \to H_*(B(G/O); Z/2)$$

maps onto the generators of $H_*(B(G/O); Z/2)$. The image of σ_* consists of primitive elements, so, as a Hopf algebra $H_*(B(G/O); Z/2)$ is primitively generated.

The subpolynomial algebra $P\{y_I \mid \ell(I) > 2\}$, where $\ell(I)$ is the number of entries in I, will play only a minor and isolated role in our calculations. It is annihilated by the map

$$B(r\times s): B(G/O) \to B^2O \times B(G/TOP)$$

and survives as a polynomial algebra in $H_*(BTOP; \mathbf{Z}/2)$. The non-trivial phenomena take place in the subalgebra $P\{\cdots, f_{2i+1}, \cdots\} \otimes P\{\cdots, y_{i,j}, \cdots\}$. This is closed under the action of the Steenrod algebra by 6.11 and because the suspension σ_* commutes with the Sq^i_*. In particular we note that

12.5
$$Sq_*^{2a+1}(f_{2i+1}) = 0$$
$$Sq_*^{2a}(f_{2i+1}) = \binom{i-a}{a} f_{2i-2a+1}\,.$$

Let $\phi_{2i+1} \in H^{2i+1}(B(G/O); \mathbf{Z}/2)$ be dual to f_{2i+1} in the monomial basis given in 12.4. From 12.5 we have

LEMMA 12.6. *The* ϕ_{2i+1} *are contained in the submodule over the Steenrod algebra with generators* ϕ_{2^i-1}. *Moreover,*

$$Sq^{2^j-2}(\phi_{2^j-1}) = \phi_{2^{j+1}-3}$$

and the ϕ_{4i+1} *are contained in the submodule over the Steenrod algebra with generators* ϕ_{2^j-3}.

REMARK 12.7. Recall from Chapter 3.D the detecting subgroup $V_2 \xrightarrow{I} \Sigma_4$, and consider the composition

$$BV_2 \xrightarrow{BI} B\Sigma_4 \xrightarrow{\theta_4} SG \xrightarrow{\pi} G/O$$

where $\theta_4 = i_4 * [-3]$. We note that π_* is injective on the image of $(\theta_4 \circ BI)_*$ and that its image is the subgroup spanned by the elements e_{2i}

and $x_{i,j}$. So, the most interesting part of $H^*(G/O; \mathbf{Z}/2)$ is detected on BV_2.

We now review the necessary $\mathbf{Z}_{(2)}$ *integral information.*

Rationally, the space $B(G/TOP)$ is a product of Eilenberg-MacLane spaces and its rational cohomology is an exterior algebra on generators of dimension $4i+1$. Since TOP/O has finite homotopy groups $B(G/O) \to B(G/TOP)$ is a rational equivalence, so

12.8 $$H^*(B(G/O); \mathbf{Q}) = E\{\bar{\phi}_5, \bar{\phi}_9, \cdots, \bar{\phi}_{4i+1}, \cdots\}$$

where each $\bar{\phi}_{4i+1}$ is a primitive generator. The suspension σ^* passing from the rational cohomology of $B(G/O)$ to the rational cohomology of G/O maps the primitive generators monomorphically.

Let $\alpha: BSO \to G/O[2]$ be a solution of the Adams conjecture as in 5.13 (with $k=3$, say) and let $y_{4n} \in F_*(BSO; \mathbf{Z}_{(2)})$ be the generator in the image of the canonical line bundle $H: CP^\infty \to BSO$, that is, the generator dual to p_4^n. We now fix the primitive generators $\bar{\phi}_{4i+1}$ of 12.8 by requiring

12.9 $$<\sigma^*(\bar{\phi}_{4i+1}), \alpha_*(y_{4i})> = +1\,.$$

In [74, p. 62 and p. 72] it was shown that primitive elements of $H^{4i+1}(B(G/O); \mathbf{Z}_{(2)})$ are detected by their rational and $\mathbf{Z}/2$ reductions. Moreover, from [74, p. 74] we have

LEMMA 12.10. (a) *Reduction gives a monomorphism*

$$\rho: PH^{4i+1}(B(G/O); \mathbf{Z}_{(2)}) \to PH^{4i+1}(B(G/O); \mathbf{Q}) \oplus PH^{4i+1}(B(G/O); \mathbf{Z}/2)\,.$$

(b) *The pair* $(\bar{\phi}_{4i+1}, \phi_{4i+1})$ *from* 12.6 *and* 12.9 *is in the image of* ρ.

We define $\hat{\phi}_{4i+1} \in PH^{4i+1}(B(G/O); \mathbf{Z}_{(2)})$ by

12.11 $$\rho(\hat{\phi}_{4i+1}) = \bar{\phi}_{4i+1} \oplus \phi_{4i+1}\,.$$

In Chapter 7 we proved that $B^2(G/TOP)[2]$ splits as a product of Eilenberg-MacLane spaces. In particular there are primitive fundamental classes

12.12 $\quad k_{4i-1} \in PH^{4i-1}(B(G/TOP); Z/2), k_{4i+1} \in PH^{4i+1}(B(G/TOP); Z_{(2)})$

but there are many possible choices for such classes. We must specify the k_{4i-1} and k_{4i+1} so that they are properly connected to the universal surgery classes of $H^{4i-2}(G/TOP; Z/2)$ and $H^{4i}(G/TOP; Z_{(2)})$, cf. 4.9 and 4.32.

This is easy for the k_{4i-1}. Since $G/TOP[2]$ is a product of Eilenberg-MacLane spaces it follows from 7.2 that

$$\sigma^*: PH^{4i-1}(B(G/TOP); Z/2) \to PH^{4i-2}(G/TOP; Z/2)$$

is an isomorphism. We choose k_{4i-1} to be the unique class with $\sigma^*(k_{4i-1}) = K_{4i-2}$, where K_{4i-2} is the class defined in 4.9.

For k_{4i+1} things are harder, since the suspension homomorphism

$$\sigma^*: PH^{4i+1}(B(G/TOP); Z_{(2)}) \to PH^{4i}(G/TOP; Z_{(2)})$$

is not surjective. Let $K_{4i} \in H^{4i}(G/TOP; Z_{(2)})$ be the universal surgery class from [91], and let

12.13 $$k_{4i} = \frac{1}{8i} s_i(8K_4, 8K_8, \cdots, 8K_{4i})$$

be its primitive form (see Chapter 9.C). It is an open question if k_{4i} itself desuspends. However, in [78, p. 299] we proved that the image of the composition

$$\text{Torsion } PH^*(G/TOP; Z_{(2)}) \longrightarrow PH^*(G/TOP; Z/2) \xrightarrow{s^*} PH^*(G/O; Z/2)$$

is contained in $Sq^1 \text{Image}(s^*)$.

The double suspensions

$$\sigma^* \circ \sigma^*: QH^{4i+2}(B^2(G/TOP); Q) \to PH^{4i}(G/TOP; Q)$$
$$\sigma^* \circ \sigma^*: QH^{4i+2}(B^2(G/TOP); Z/2) \to PH^{4i}(G/TOP; Z/2)$$

are isomorphisms, and it follows that there exists a fundamental class $k_{4i+2} \in H^{4i+2}(B^2(G/TOP), Z_{(2)})$ with $\sigma^* \circ \sigma^*(k_{4i+2}) \otimes Q = k_{4i} \otimes Q$ and such that the $Z/2$ reduction of $\sigma^* \circ \sigma^*(k_{4i+2}) - k_{4i}$ maps to zero in $H^*(G/O; Z/2)$. We then set $k_{4i+1} = \sigma^*(k_{4i+2})$. Note from 12.4 and 12.10 that $(Bs)^*(k_{4i+1})$ is a well-defined element of $PH^{4i+1}(B(G/O); Z_{(2)})$.

With these choices of k_{4i-1}, k_{4i+1} and $\hat{\phi}_{4i+1}$ we have the following main result

THEOREM 12.14. *The map* $Bs: B(G/O) \to B(G/TOP)[2]$ *is characterized by the properties*:

(i) $(Bs)^*(k_{4i+1}) = 2^{\alpha(i)-1} u_i \hat{\phi}_{4i+1}$

(ii) $(Bs)^*(k_{4i-1}) = 0$ *if* $\alpha(i) > 1$

(iii) $(Bs)^*(k_{4i-1}) = \phi_{4i-1}$ *if* $\alpha(i) = 1$

where $\alpha(i)$ *is the number of non-zero terms in the dyadic expansion of* i *and* u_i *a unit of* $Z_{(2)}$.

(The actual multiple of $\hat{\phi}_{4i+1}$ in (i) above is $\frac{1}{(2i)!} 2^{2i-1}(2^{2i-1}-1)(3^{2i}-1) B_{2i}/4i$ which has 2-adic valuation equal to $\alpha(i)-1$.)

Proof. By 12.10 it suffices to check with rational coefficients and with $Z/2$ coefficients. The rational calculation follows from 4.22 and 5.13 since the primitive series for the $\mathcal{L}$-genus is

$$P(\mathcal{L}) = \sum (-1)^n \left[\frac{2^{2n+1}(2^{2n-1}-1)}{(2n-1)!}\right] B_{2n}/4n \; z^n$$

and since the suspension passing from $PH^{4i+1}(B(G/O); Q)$ to $PH^{4i}(G/O; Q)$ is injective.

With $Z/2$ coefficient the result essentially follows from [35, p. 134]. (The only difficulty is to see that [35, (9.1)(iii)] implies that $s^*(k_{2^{j+1}-2})$

is dual to $e_{2^{j+1}-2}$. But this follows since in $H_*(QS^0; Z/2)$ we have for all $s, t \geq 1$ with $s+t = 2^{i+1}-2$ the equation

$$(*) \quad Q^s[1] * Q^t[1] * [-3] = Q^{2^i-1}Q^{2^i-1}[1] * [-3] + \sum a_\nu (Q^{I_\nu}[1] * [-2^{\ell(I_\nu)}+1])$$

(modulo decomposable elements in the composition product). The equation (*) is a (non-trivial) exercise in mod.2 binomial arithmetic, using 6.8, 6.11 and 6.18, but we leave the details to the reader.

Finally, we must calculate the induced map

$$(Br)^* : H^*(B^2O; Z_{(2)}) \to H^*(B(G/O); Z_{(2)}) .$$

First, recall that

$$H^*(B^2O; Z/2) = E\{v_2, v_3, \cdots\}$$

with Bockstein relations $Sq^1(v_{2i}) = v_{2i+1}$. Therefore, in the Bockstein spectral sequence we have

$$E_2 = E_\infty = E\{v_2v_3, \cdots, v_{2i}v_{2i+1}, \cdots\} .$$

The 2-torsion is all of order 2 and the free part $F^*(B^2O; Z_{(2)})$ is an exterior algebra in primitive classes h_{4i+1}. The next result completely specifies $(Br)^*$.

THEOREM 12.15. $(Br)^*$ *is zero with* $Z/2$ *coefficients and* $(Br)^*(h_{4i+1}) = (3^{2i}-1)\hat{\phi}_{4i+1}$.

Proof. Consider the fibration sequence

$$BO \longrightarrow BG \xrightarrow{p} B(G/O) \xrightarrow{Br} B^2O .$$

Since p_* is surjective with $Z/2$ coefficients $(Br)_*$ is the zero map, and we are left with the rational calculation. But here one uses (as in the proof of 12.14) that the composition

$$BO \xrightarrow{\alpha} G/O[2] \xrightarrow{r} BO$$

is ψ^3-1. This completes the proof.

REMARK 12.16. In 5.18 we gave the splitting of $G/O[2]$,

$$\alpha \times i : BSO[2] \times \mathrm{cok}\, J_2 \xrightarrow{\simeq} G/O[2]$$

where $\mathrm{cok}\, J_2$ is the fiber of $e : G/O[2] \to BSO^{\otimes}[2]$ and $BSO[2]$ is mapped into $G/O[2]$ by a solution of the Adams conjecture, cf. 5.13. It is of interest to compare the natural map $s : G/O[2] \to G/TOP[2]$ with this splitting.

As above, let $K_{4n} \in H^{4n}(G/TOP; \mathbf{Z}_{(2)})$ be the class from [91] and let $\tilde{K}_{4n} \in H^{4n}(G/TOP; \mathbf{Z}_{(2)})$ be the class from [103]. Their difference was calculated in [34],

$$\sum \tilde{K}_{4n} - \sum K_{4n} = \beta\left(\sum_{i=0}^{\infty} Sq^{2^i}\right) Sq^1\left(\sum_{n=1}^{\infty} K_{4n-2}\right),$$

where K_{4n-2} is the class in 4.9. It is surprising that $i^*s^*(K_{4n}) \neq 0$ in $H^{4n}(\mathrm{cok}\, J_2; \mathbf{Z}_{(2)})$ when $\alpha(n) = 2$. In contrast $i^*s^*(\tilde{K}_{4n}) = 0$ for all $n \geq 1$ and in fact from [77, p. 191], we have

12.17 $$s^*(\tilde{K}_{4n}) = e^* \circ (\gamma_\alpha^{-1})^*(ph_{1,n-1})$$

where $\gamma_\alpha : BSO^{\oplus}[2] \to BSO^{\otimes}[2]$ is the H-equivalence also considered in Chapter 9.B, and $ph_{1,n-1} \in H^{4n}(BSO; \mathbf{Z}_{(2)})$ is the class whose rational reduction is $2^{2n-2}ph_{4n}$ and whose $\mathbf{Z}/2$ reduction is $\chi(Sq^{4n-4})(p_4)$. Note also from 12.17 that

$$\alpha^*s^*(\tilde{K}_{4n}) = ph_{1,n-1}\,.$$

For a particular solution of the Adams conjecture it was proved in [34] that

$$\alpha^*s^*(K_{2^r-2}) = \sum_{u,v} \sum_{i=0}^{2^r} t_u^i\, t_v^{2^r-i}$$

where we consider $H^*(BO; \mathbf{Z}/2)$ as the symmetric functions in the variables $t_1, t_2, \cdots$, cf. Chapter 1.B. Finally we remark that $i^*s^*(K_{2^r-2}) \neq 0$ in

$H^*(\mathrm{cok}\ J_2; \mathbf{Z}/2)$, and is spherical if and only if there is a stable homotopy class in $\pi^s_{2^r-2}(S^0)$ with Arf invariant one.

B. *The Leray-Serre spectral sequence for* BTOP

We consider 3 elementary model spectral sequences

12.18
$$E^2_{*,*} = E\{f\} \otimes E\{z\}, \quad \deg(f) = 2n+1, \quad \deg(z) = 2n$$
$$d_{2n+1}(f) = z .$$

Then $E^{2n+2} = E^\infty = E\{fz\}$ and the picture of 12.18 is

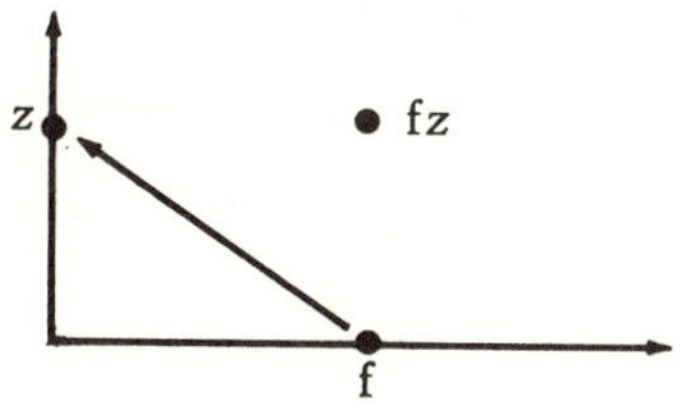

The next model is

12.19
$$E^2_{*,*} = P\{y\} \otimes E\{z\}, \quad \deg(y) = i+2j+1, \quad \deg(z) = i+2j$$
$$d_{2j+i+1}(y) = z$$

with picture

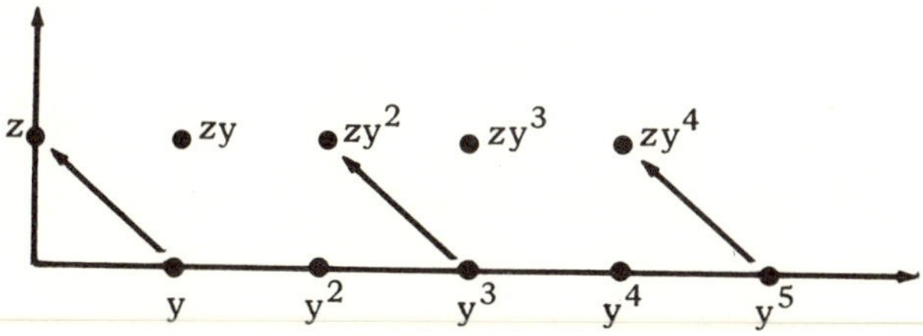

Here $E^{2j+i+2} = E^\infty = E\{yz\} \otimes P\{y^2\}$. Finally we have

$$E^2_{*,*} = \Gamma\{y\} \otimes E\{z\}, \quad \deg(y) = r, \quad \deg(z) = r-1$$

12.20

$$d_r(y) = z, \quad d_r(\gamma_i(y)) = \gamma_{i-1}(y)\, z .$$

Here $E^\infty_{*,*} = 0$ and the picture is

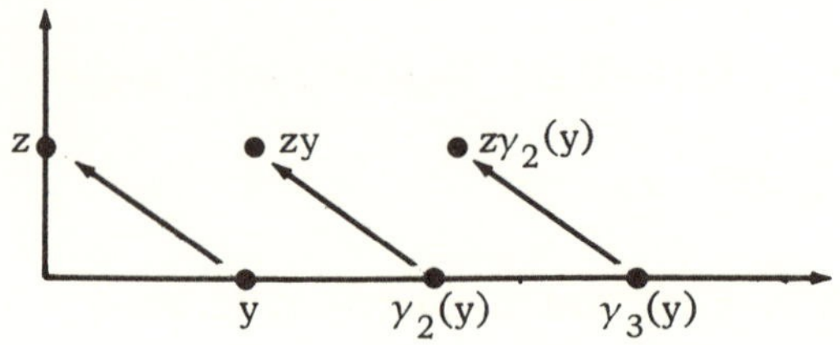

THEOREM 12.21. *The Leray-Serre spectral sequence for the fibering* $BO \times G/TOP \to BTOP \to B(G/O)$ *is a tensor product of the two models* 12.18, 12.19 *above and models with no differentials. Hence its* E^∞*-term has the form*

$$E^\infty = H_*(BO; \mathbf{Z}/2) \otimes P\{y^2_{i,j} \mid 0 < i \leq j\} \otimes P\{y_I \mid \text{length } I \geq 3\}$$

$$\otimes E\{\cdots, f_{2i+1} d(f_{2i+1}), \cdots, y_{i,j} d(y_{i,j}), \cdots\} \otimes L$$

where L *is the quotient of* $H_*(G/TOP; \mathbf{Z}/2)$,

$$L = H_*(G/TOP; \mathbf{Z}/2) // E\{\cdots, s_*(e_{2i}), \cdots, s_*(x_{i,j}), \cdots\} .$$

Proof. $(E^r_{*,*}, d^r)$ is a spectral sequence of Hopf algebras and the transgressive differentials are determined by $d^r(y_I) = \phi_*(x_I)$ where ϕ is the composition

$$\phi : G/O \longrightarrow BO \times G/TOP \xrightarrow{1 \times \chi} BO \times G/TOP .$$

This follows upon comparison with the spectral sequence passing from $H_*(G/O; \mathbf{Z}/2)$ to $H_*(B(G/O); \mathbf{Z}/2)$. Now the lowest differential d_r certainly has a primitive image and so must be transgressive. Moreover, since $\text{Im}(\phi_*) \subset H_*(G/TOP)$ the the image of the differential is an exterior generator. Thus we obtain one or the other of our two models and can write $E_r = M \otimes P$ where P has trivial differential and M is a tensor product of these models. But now note for filtration reasons that zy or zf must be an infinite cycle and no differentials can hit it, again for filtration reasons so it survives to E^∞ and we can write $E_{r+1} = E\{zf, \cdots\} \otimes Q$ where Q is a tensor product $E\{\cdots, f, \cdots\} \otimes P\{\cdots, y_I, \cdots\}$ from the base with $H_*(BO; \mathbf{Z}/2) \otimes G_{r+1}$ from the fiber. We can now repeat the argument above and 12.21 follows.

The above result gives the additive structure of $H_*(BTOP; \mathbf{Z}/2)$ and most of the ring structure. However, to obtain the full ring structure we must argue that the elements $f_{2i+1} d(f_{2i+1})$ and $y_{i,j} d(y_{i,j})$ which are exterior in E^∞ can be represented by exterior elements in $H_*(BTOP; \mathbf{Z}/2)$. To this end it is more convenient to look at the Leray-Serre spectral sequence of the second fibering in 12.1,

$$BTOP \to B(G/O) \to B(G/TOP) \times B^2O .$$

In fact, we can disregard the factor B^2O and instead look at the fibering

12.22 $$B(TOP/O) \to B(G/O) \to B(G/TOP)$$

since it is easily seen that

$$H_*(BTOP; \mathbf{Z}/2) \cong H_*(BO; \mathbf{Z}/2) \otimes H_*(B(TOP/O); \mathbf{Z}/2) .$$

But $B(G/TOP)[2]$ is a product of Eilenberg-MacLane spaces so $H^*(B(G/TOP); \mathbf{Z}/2)$ is a polynomial algebra on primitive generators. Dually, $H_*(B(G/TOP); \mathbf{Z}/2)$ is a divided power algebra,

12.23 $$H_*(B(G/TOP); \mathbf{Z}/2) = \Gamma\{\cdots, (Sq^I k_{2m+1})^*, \cdots\} .$$

Among the generators in 12.23 we have the elements $\gamma((Bs_*(f_{2i+1}))$ and $\gamma(Bs_*(y_{i,j}))$ and these transgress in the Leray-Serre spectral sequence of 12.22 to elements "$f_{2i+1} df_{2i+1}$" and "$y_{i,j} dy_{i,j}$" in $H_*(B(TOP/O); \mathbf{Z}/2)$ which represent $f_{2i+1} df_{2i+1}$ and $y_{i,j} dy_{i,j}$. But then, using the coalgebra structure, or equivalently, by dualizing to the cohomology spectral sequence of 12.22,

12.24
$$d_{4i+2}(\gamma^2((Bs)_*(f_{2i+1}))) = \text{``}f_{2i+1}df_{2i+1}\text{''} \cdot \gamma(Bs_*(f_{2i+1}))$$
$$d_{2i+4j}(\gamma^2((Bs)_*(y_{i,j}))) = \text{``}y_{i,j} \cdot dy_{i,j}\text{''} \cdot \gamma(Bs_*(y_{i,j})) .$$

In particular we have

$$d_{4i+2}(\text{``}f_{2i+1} \cdot df_{2i+1}\text{''} \cdot \gamma(Bs_*(f_{2i+1}))) = 0$$

$$d_{2i+4j}(\text{``}y_{i,j}dy_{i,j}\text{''} \cdot \gamma(Bs_*(y_{i,j}))) = 0$$

and hence

$$(\text{``}f_{2i+1}df_{2i+1}\text{''})^2 = 0 \quad (\text{``}y_{i,j}dy_{i,j}\text{''})^2 = 0 .$$

We have proved (see also [35]).

THEOREM 12.25. $H_*(BTOP; Z/2)$ *is isomorphic as an algebra to the* E^∞ *described in* 12.21.

REMARK 12.26. It is not hard to carry the discussion further so as to show that the Leray-Serre spectral sequence of 12.22 is a tensor product of the models 12.20 and models with no differentials.

REMARK 12.27. The description of $H_*(BPL; Z/2)$ is analogous but slightly more complicated. The difference arises since $H_*(G/PL; Z/2)$ is no longer an exterior algebra. Indeed, we have $s_*(e_2)^2 \neq 0$ as one can see from the footnote to Theorem 4.32. The action of the Steenrod algebra generates further exotic products of the form $s_*(e_{2i})^2 \neq 0$ and $s_*(x_{i,j})^2 \neq 0$ but all 4'th powers vanish. For such exotic products the associated terms fdf and ydy in 12.21 would not appear for BPL. Further, of course, the L-factor would be smaller. We leave the details to the reader, who might also consult [35].

CHAPTER 13

THE BOCKSTEIN SPECTRAL SEQUENCE FOR BTOP

In this chapter we complete the 2-local analysis of BTOP by calculating its mod.2 Bockstein spectral sequence. The results imply a calculation of $F_*(\mathrm{BTOP}) \otimes \mathbf{Z}_{(2)}$ and $\Omega_*^{\mathrm{TOP}}/\mathrm{Tor}$ as described in Chapter 8 and Chapter 10.C. In principle, they also give the 2-torsion structure of the topological (and PL) cobordism groups in dimensions $\neq 4$. This is exploited in Chapter 14.

The calculation is based on various 'universal Bockstein relations,' the proof of which are given in the Appendix.

We refer the reader to [23], [74] for the definition and standard properties of the Bockstein spectral sequence. Here it will be denoted (E_*^r, ∂_r) to avoid confusions with the Leray-Serre sequence. The E^1-term is $H_*(X; \mathbf{Z}/2)$ and the E^∞-term is $F_*(X) \otimes \mathbf{Z}/2$.

A. *The Bockstein spectral sequences for* BO, G/TOP *and* B(G/O)

LEMMA 13.1. *In* $E_*^1(\mathrm{BO})$, $\partial_1(x_{2i}) = x_{2i-1}$ *and* $E_*^2 = E_*^\infty = P\{x_2^2, \cdots, x_{2n}^2, \cdots\}$.

To give the Bockstein sequence for G/TOP we need two models. The first we write

13.2 $$(E\{x\} \otimes \Gamma\{y\}, r)$$

where deg x is odd, deg y is even and $\partial_r(y) = x$. Here $\Gamma\{y\}$ is the divided power algebra on y tensored with $\mathbf{Z}/2$, that is, the Hopf algebra dual to $P\{y^*\}$. Consequently, as an algebra

$$\Gamma\{y\} = E\{y\} \otimes E\{\gamma(y)\} \otimes \cdots \otimes E\{\gamma^j(y)\} \otimes \cdots .$$

The differentials are given by

$$\partial_r(\gamma^j(y)) = xy \cdot \gamma(y) \cdots \gamma^{j-1}(y) .$$

So $E_*^{r+1} = E_*^\infty = Z/2$.

The second model we write

13.3 $$(\Gamma\{x\} \otimes E\{y\}, r)$$

where $\deg x$ is even, $\deg y$ is odd and $\partial_r(y) = x$. Then $E_*^{r+1} = (\Gamma\{\gamma(x)\} \otimes E\{xy\}, r+1)$ and the model keeps replicating itself,

$$E_*^{r+j} = (\Gamma\{\gamma^j(x)\} \otimes E\{yx\gamma(x) \cdots \gamma^{j-1}(x)\}, r+j) .$$

We have the following well-known lemma (see e.g. [23])

LEMMA 13.4. a) *The* B.S.S. *for* $K(Z, 2n)$ *is*

$$\Gamma\{\iota_{2n}\} \otimes \bigotimes_I (E\{x_I\} \otimes \Gamma\{y_I\}, 1) \otimes \bigotimes_J (\Gamma\{x_J\} \otimes E\{y_J\}, 1)$$

for suitable indexing sets I, J.

b) *The* B.S.S. *for* $K(Z/2, 2n)$ *is*

$$\bigotimes_I (E\{x_I\} \otimes \Gamma\{y_I\}, 1) \otimes \bigotimes_J (\Gamma\{x_J\} \otimes E\{y_J\}, 1) .$$

In the first case $E_*^\infty = \Gamma\{\iota_{2n}\}$ *and in the second case* $E_*^\infty = Z/2$.

To give the Bockstein sequence for $B(G/O)$ we need 3 types of models. The first is

13.5
$$M_i = E\{f_{4i+3}\} \otimes P\{y_{1,2i+1}\}$$
$$\partial_1(y_{1,2i+1}) = f_{4i+3} .$$

Then $E_*^2 = E\{fy\} \otimes P\{y^2\}$ and $\partial_2(y^2) = fy$. In general $E_*^r = E\{fy^{2^{r-1}-1}\} \otimes P\{y^{2^{r-1}}\}$ with $\partial_r(y^{2^{r-1}}) = fy^{2^{r-1}-1}$.

The second model is

13.6 $$N_I = P\{y_I, y_{I_1}, y_{(2,I_1)}, \cdots, y_{(2,\cdots,2,I_1)}, \cdots\}$$

with $\partial_1(y_{I_1}) = y_I, \partial_1(y_{(2,I_1)}) = y_{I_1}^2, \partial_1(y_{(2,2,I_1)}) = y_{(2,I_1)}^2$ etc. Then $E_*^2 = E_*^\infty = Z/2$.

Here we take the sequences $I = (i_1, \cdots, i_r)$ with

13.7 $$i_1 \equiv 1 \pmod 2 \quad \text{and} \quad i_2 \equiv 0 \pmod 2 ,$$

and we have $I_1 = I + (1, 0, \cdots, 0)$.

The third model is

13.8 $$Q_I = P\{y_I, y_{I_1}, y_{(2,I)}, y_{(2,2,I)}, \cdots, y_{(2,\cdots,2,I)}, \cdots\}$$

with $\partial_1(y_{I_1}) = y_I, \partial_1(y_{(2,I)}) = y_I^2, \partial_1(y_{(2,2,I)}) = y_{(2,I)}^2$ etc. Then

$$E_*^2 = E\{y_I y_{I_1} + y_{(2,I)}\} \otimes P\{y_{I_1}^2\}, \partial_2(y_{I_1}^2) = y_I y_{I_1} + y_{(2,I)} = z_{I_1}$$

and in general $E_*^{r+1} = E\{zy^{2r-1}\} \otimes P\{y^{2^r}\}$ with $\partial_{r+1}(y^{2^r}) = zy^{2^{r-1}}$, where $z = z_{I_1}$; $y = y_{I_1}$. Here the conditions on I are

13.9 $$i_1 \equiv 0 \pmod 2 \quad \text{and} \quad i_2 \equiv 1 \pmod 2$$

The final model we need is simply $E\{f_{4i+1}\}$ with trivial differentials. From [74, p. 72] we have

THEOREM 13.10.

$$E^1(B(G/O)) = E\{f_5, \cdots, f_{4n+1}, \cdots\} \otimes \bigotimes_{i \geq 1} M_i \otimes N_I \otimes Q_J$$

where $i \geq 1$, I, *and* J *run over sequences satisfying* 13.7 *and* 13.9, *and the differentials are as described above.*

B. *The spectral sequence for* BTOP

The starting term of the spectral sequence is given in Theorems 12.21 and 12.25. For convenience we let $h_{0,i} \in H_{4i+3}(\mathrm{BTOP}; \mathbf{Z}/2)$ and $h_{i,j} \in H_{2^{i+4}j+3}(\mathrm{BTOP}; \mathbf{Z}/2)$ be exterior generators which reduce to

13.11 $$h_{0,i} = f_{2i+1}\, df_{2i+1}\,, \quad h_{i,j} = y_{i,j}\, d(y_{i,j})$$

in the E^{∞} term given in 12.21. (This, of course, does not completely specify the elements; the indeterminacy consists of exterior terms of lower filtration in the Leray-Serre spectral sequence used in 12.21.)

We first determine the ∂_1 differential. On most of $H_*(\mathrm{BTOP}; \mathbf{Z}/2)$ it is determined by what happens in $\mathrm{BO} \times G/\mathrm{TOP}$ or $B(G/O)$. The places where things change involve the $h_{i,j}$ and the $y_{i,j}^2$.

LEMMA 13.12. a) *Let* i *be odd and* j *even. Then*

$$\partial_1(y_{i,j}^2) = h_{i,j} \quad \text{and} \quad \partial_1(h_{i+1,j}) = J_{i,j}$$

where $J_{i,j} \in \mathrm{Im}(\hat{j}_*: H_*(G/\mathrm{TOP}; \mathbf{Z}/2) \to H_*(\mathrm{BTOP}; \mathbf{Z}/2))$, *say* $J_{i,j} = \hat{j}_*(\tilde{J}_{i,j})$, *and* $\partial_1(\tilde{J}_{i,j})$ *is a non-primitive decomposable in the ideal generated by the elements* $s_*(e_{2^i})$ *and* $s_*(x_{i,j})$.

b) *Let* i *be even and* j *odd. Then*

$$\partial_1(h_{i,j}) = L_{i,j} \qquad \text{if} \quad i > 0$$

$$\partial_1(y_{i+1,j}^2) = h_{i+1,j} \qquad \text{if} \quad i \geq 0\,,$$

where $L_{i,j} = \hat{j}_*(\tilde{L}_{i,j})$ *and* $\psi(\tilde{L}_{i,j}) = s_*(x_{i,j}) \otimes s_*(x_{i,j})$ *with* $\tilde{L}_{i,j}$ *surviving to* E_*^2.

LEMMA 13.13. $\partial_1(h_{0,j}) = 0$ *if* $j \neq 2^i$ *and* $\partial_1(h_{(0,2^i)}) = K_{2^{i+2}}$, *where* $K_{2^{i+2}}$ *is the image of the spherical generator of* $H_{2^{i+2}}(G/\mathrm{TOP}; \mathbf{Z}/2)$.

We defer the proofs of 13.12 and 13.13 to the Appendix.

COROLLARY 13.14.*)

$$E^2_*(\mathrm{BTOP}) = \bigotimes_{i\ odd} P\{y^4_{i,j}\} \otimes E\{h_{i,j} y^2_{i,j}\}$$

$$\otimes P\{y_I^2 \mid i_1, i_2\ \mathit{odd},\ \mathit{length}\ I \geq 3\} \otimes E\{z_I \mid i_1, i_2\ \mathit{odd},\ \mathit{length}\ I \geq 3\}$$

$$\otimes \Gamma\{\gamma(L_{i,j}) \mid i > 0,\ \mathit{even},\ j\ \mathit{odd}\} \otimes E\{L_{i,j} h_{i,j} \mid i > 0\ \mathit{even},\ j\ \mathit{odd}\}$$

$$\otimes \Gamma\{\gamma(J_{i,j}) \mid i\ \mathit{odd},\ j\ \mathit{even}\} \otimes E\{J_{i,j} h_{i+1,j} \mid i\ \mathit{odd},\ j\ \mathit{even}\}$$

$$\otimes \Gamma\{\gamma(K_{2^{r+2}}) \mid r \geq 0\} \otimes E\{K_{2^{r+2}} h_{0,2^r} \mid r \geq 0\}$$

$$\otimes \bigotimes_R \Gamma\{V_R\} \otimes E\{W_R\}$$

$$\otimes P\{x_2^2, \cdots, x_{2i}^2, \cdots\} \otimes \Gamma\{K_{4i} \mid i \neq 2^r\} \otimes E\{h_{0,j} \mid j \neq 2^r\}\ .$$

In 13.14 the V_R come from $H_*(G/TOP; Z/2)$ and have dimensions $\equiv 0(4)$. Also the W_R come from $H_*(G/TOP; Z/2)$ and $\partial_2(W_R) = V_R$. The elements x_{2i}^2 and K_{4i} come from $E^2_*(BO \times G/TOP)$ where they represent infinite cycles surviving to E^∞_*. Hence, the same must be true of the images, which, though there may be differentials hitting them, cannot themselves have any non-zero differentials. Also, the z_I above are merely shorthand ways of writing the $y_{I-(1,0,\cdots)} \cdot y_I + y_{(2,I-(1,0,\cdots))}$ as in 13.8.

As far as the Hopf algebra structure of E^2 is concerned the y_I^2, $y^4_{i,j}$, $\gamma(L_{i,j})$, $\gamma(J_{i,j})$, V_R, $\gamma(K_{2^{r+2}})$, and K_{4i} can be assumed primitive. We may also assume all the exterior generators are primitive for dimensional reasons. But the $\gamma^r(V_R)$, $\gamma^{r+1}(L_{i,j})$, $\gamma^{r+1}(J_{i,j})$, $\gamma^{r+1}(K_{2^{i+2}})$ and $\gamma^r(K_{4j})$ are not primitive when $r \geq 1$.

We now turn to the higher differentials in the B.S.S. First, note that most of the higher differentials for BTOP are determined for simple universal reasons (compare with the models 13.3, and 13.8). Indeed the only problem left is to determine the higher differentials on the $h_{0,j}$ with $j \neq 2^r$.

The proof of the next lemma is given in the Appendix.

*)The classes $y_{i,j}$ x_i^2 here are not to be confused with the classes on page 197, in Theorem 10.9.

LEMMA 13.15. *There is a space* Y_{4j} *for* j *not a power of* 2 *and a mapping*

$$\lambda_j : \mathrm{BTOP} \to Y_{4j}$$

with $\lambda_j{}_*(h_{0,j})$, $\lambda_j{}_*(K_{4j})$ *and* $\lambda_j{}_*(N_{4j})$ *all non-zero and*

$$\partial_s(\lambda_j{}_*(h_{0,j})) = 0 \quad \text{if} \quad s < \min(4, \alpha(j))$$

$$\begin{aligned}\partial_r(\lambda_j{}_*(h_{0,j})) &= \lambda_j{}_*(K_{4j}) \quad \text{if} \quad \alpha(j)-4 < \nu_2(j) \\ &= \lambda_j{}_*(N_{4j}) \quad \text{if} \quad \alpha(j)-4 > \nu_2(j) \\ &= \lambda_j{}_*(K_{4j}+N_{4j}) \quad \text{if} \quad \alpha(j)-4 = \nu_2(j),\end{aligned}$$

where $r = \min(\alpha(j), \nu_2(j)+4)$ *and* $N_{4j} \in P\{x_2^2, \cdots, x_{2i}^2, \cdots\}$. *Moreover,* Y_{4j} *is the total space of a fibering, the map* λ_j *extends to a map of the first fibering in* 12.1 *and in the map of Serre spectral sequences the image of* $h_{0,j}$ *is non-trivial.*

So far the classes $h_{0,j}$ have not been specified completely. We have,

COROLLARY 13.16. *The elements* $h_{0,q}$ *can be so chosen that* $\partial_r(h_{0,q})$ *belongs to the quotient of* $\Gamma\{\cdots, K_{4i}, \cdots\} \otimes P\{\cdots, x_{2i}^2, \cdots\}$ *in* $E_*^r(\mathrm{BTOP})$.

Proof. Since $h_{0,q}$ is primitive,

$$\partial_r(h_{0,q}) = \sum \varepsilon_R \gamma^{r-2}(V_R) + \sum s_{ij}\gamma^{r-1}(J_{i,j}) + \sum t_{ij}\gamma^{r-1}(L_{i,j})$$
$$+ \theta K_{4j} + \bar{\theta} N_{4j} + \sum_{i \text{ odd}} \varepsilon_{i,j} y_{i,j}^{2^s}.$$

(The remaining $y_I^{2^s}$ all map nontrivially to $E^r(B(G/O))$ where either a differential is non-trivial on them or they survive to $E_*^{r+1}(B(G/O))$.) But

$$\partial_r(W_R V_R \cdots \gamma^{r-3}(V_R)) = \gamma^{r-2}(V_R)$$

$$\partial_r(h_{i,j} L_{i,j} \cdots \gamma^{r-2}(L_{i,j})) = \gamma^{r-1}(L_{i,j})$$

$$\partial_r(h_{i,j} J_{i,j} \cdots \gamma^{r-2}(J_{i,j})) = \gamma^{r-1}(J_{i,j})$$

so changing $h_{0,q}$ by

13.17 $$\sum \epsilon_R W_R \cdots \gamma^{r-3}(V_R) + \sum s_{ij} h_{i,j} \cdots \gamma^{r-2}(L_{i,j}) + \sum t_{ij} h_{i,j} \cdots \gamma^{r-2}(J_{i,j})$$

reduces us to

13.18 $$\partial_r(h_{0,q}) = \theta K_{4q} + \bar{\theta} N_{4q} + \sum_{i \text{ odd}} \epsilon_{ij} y_{i,j}^{2^s} .$$

(Note that the term 13.17 has strictly smaller filtration in the Leray-Serre sequence for 12.1 and so lies in the indeterminancy of $h_{0,q}$. Thus it does not affect 13.15.

Now, assume 13.16 is not true. Then there is a smallest j so it fails and we consider this differential. If in 13.18 θ or $\bar{\theta}$ is non-zero, then $r = \alpha(q)$ or $\nu_2(q) + 4$ whichever is smaller. If $r = \nu_2(q) + 4$ then all powers of the $y_{i,j}^2$ will have been used in lower differentials except $y_{i,j}^{2^{\nu_2(q)+4}}$ which have degree divisible by $2^{\nu_2(q)+5}$ contradicting that $\deg(\partial_r(h_{0,q})) = 4q$.

Hence we can assume $r = \alpha(q) < \nu_2(q) + 4$. In this case $\theta \neq 0$, $\bar{\theta} = 0$ and

13.19 $$0 = [\partial_r(h_{0,q})]^2 = \sum \epsilon_{ij} y_{i,j}^{2^{r+1+\lambda(i,j)}}$$

in E_*^{r+1} since $K_{4q}^2 = 0$. As $\lambda(i,j) \geq 1$ for all $\epsilon_{i,j} \neq 0$ we may write 13.19 as

$$\left(\sum \epsilon_{ij} y_{i,j}^{2^{r+1+\lambda(i,j)-s}}\right)^{2^s} = V^{2^s}$$

where at least one $\lambda(i,j) = s$ and V is an indecomposable in E_*^{r+1}. Now we follow the spectral sequence s steps to E_*^{r+s+1}. Here we have the primitive element $\partial_{r+1}(V) \cdot V \cdot V^2 \cdots V^{2^{s-1}}$, which for dimensional reasons must be an infinite cycle and only K_{4i} or N_{4i} can hit it. But these are known to be infinite cycles also, so it survives to E_*^∞ non-trivially. But this is a contradiction since $H_*(BTOP; Z)/Tor$ has no odd dimensional components.

It remains to consider the case where $\theta = \bar{\theta} = 0$. But here we obtain as in 13.19

$$\partial_r(h_{0,q}) = \left[\sum \varepsilon_{ij} y_{i,j}^{2^{r+\lambda(i,j)-s}}\right]^{2^s} = V^{2^s}$$

where $\lambda(i,j) \geq 1$ for all i,j and $s = \lambda(i,j)$ for at least one pair, and here again the argument proceeds as above. This completes the proof.

We now state the main result of this section

THEOREM 13.20. *In the Bockstein spectral sequence for* BTOP *(at the prime* 2*) the map*

$$E^\infty(BO \times G/TOP) \to E^\infty(BTOP)$$

is a surjection.

Proof. From 13.16, 13.14 and the remarks following 13.14 we see that

$$P\{y_{i,j}^4\} \otimes E\{h_{i,j} y_{i,j}^2\} \otimes P\{y_I^2\} \otimes E\{Z_I\} \otimes$$

$$\Gamma\{\gamma(L_{i,j}), \gamma(J_{i,j}), V_R\} \otimes E\{h_{i,j} L_{i,j}, h_{i,j} J_{i,j}, W_R\}$$

form a sub-spectral sequence converging to $Z/2$ and

13.21 $$\bigotimes \Gamma\{K_{4i}\} \otimes P\{x_{2i}^2\} \otimes E\{h_{0,j}\}$$

forms a second sub-spectral sequence converging to $H_*(BTOP)/Tor \otimes Z/2$. Moreover, 13.15 implies that each $h_{0,j}$ ultimately has a differential on it which is non-trivial. (This statement will be verified in detail in part C below.) Hence $E^\infty \subset \Gamma\{\cdots, K_{4i}, \cdots\} \otimes P\{\cdots, x_{2i}^2, \cdots\}$ which is the theorem.

C. *The differentials in the subsequence* 13.21

It is useful for studying the torsion in $\Omega_*^{TOP}(pt)$ to exactly determine the differentials in 13.21. The calculation is (not surprisingly) very close to that in the proof of 10.9.

DEFINITION 13.22. Let A_n for n odd be the DG-algebra over $Z_{(2)}$

$$A_n = \bigotimes_{i \geq 0} P\{q_{n,i}\} \otimes \Gamma\{K_{n,i}\} \otimes E\{\varepsilon_{n,i}\}$$

with derivation inductively determined by

$$\partial(\varepsilon_{n,i}) = 2^{\alpha(n)} K_{n,i} - 2^4(2^i q_{n,i} + \cdots + q_{n,0}^{2^i}) .$$

(Here $q_{n,i}$ and $K_{n,i}$ both have dimension $2^{i+2}n$.)

We calculate the Bockstein spectral sequence for A_n as follows

PROPOSITION 13.23. *If* $\alpha(n) < 4$ *then*

$$E_*^s = P\{q_{n,0}, q_{n,1}, \cdots\} \otimes E_{(1)}^s$$

where

$$E_{(1)}^1 = E_{(1)}^{\alpha(n)} = \Gamma\{K_{n,0}, K_{n,1}, \cdots\} \otimes E\{\varepsilon_{n,0}, \varepsilon_{n,1}, \cdots\}$$

$$\partial_{\alpha(n)}(\varepsilon_{n,i}) = K_{n,i} .$$

Proof. We can write

$$\partial\varepsilon_{n,i} = 2^{\alpha(n)}\{K_{n,i} - 2 \cdot 2^{3-\alpha(n)}(2^i q_{n,i} + \cdots + q_{n,0}^{2^i})\}$$

but the elements in the brackets can be taken as new divided power generators replacing the $K_{n,i}$.

PROPOSITION 13.24. *If* $\alpha(n) \geq 4$ *then*

a) *if* $s+4 \leq \alpha(n)$:

$$E^{s+4} = \Gamma\{K_{n,0}, \cdots, K_{n,\alpha(n)-4}\} \otimes \Gamma\{\overline{K}_{n,\alpha(n)-3}, \overline{K}_{n,\alpha(n)-2}, \cdots\} \otimes$$

$$P\{\overline{q}_{n,s}, \overline{q}_{n,s+1}, \cdots\} \otimes E\{\overline{\varepsilon}_{n,s}, \overline{\varepsilon}_{n,s+1}, \cdots\}.$$

b) *if* $s+4 > \alpha(n)$:

$$E^{s+4} = \Gamma\{K_{n,0}, \cdots, K_{n,\alpha(n)-4}\} \otimes P\{q_{n,\alpha(n)-3}, q_{n,\alpha(n)-2}, \cdots\} \otimes$$

$$\otimes \Gamma\{\gamma^{s+4-\alpha(n)}(K_{n,\alpha(n)-3}), \gamma^{s+4-\alpha(n)}(K_{n,\alpha(n)-2}), \cdots\} \otimes$$

$$E\{\varepsilon^{(s)}_{n,\alpha(n)-3}, \varepsilon^{(s)}_{n,\alpha(n)-2}, \cdots\}.$$

Moreover, in case (a) $\partial_{s+4}(\overline{\varepsilon}_{n,s}) = \overline{q}_{n,s}$ *and* $\partial_{\alpha(n)}(\overline{\varepsilon}_{n,\alpha(n)-4+j}) = \overline{K}_{n,\alpha(n)-4+j}$. *In case* (b) $\partial_{s+4}(\varepsilon^{(s)}_{n,\alpha(n)+i}) = \gamma^{s+4-\alpha(n)}(\overline{K}_{n,\alpha(n)+i})$.

Proof. Note first

$$\partial\varepsilon_{n,0} = 2^{\alpha(n)}K_{n,0} - 2^4 q_{n,0}$$

$$\partial(\varepsilon_{n,1} - q_{n,0}\,\varepsilon_{n,0}) = 2^{\alpha(n)}(K_{n,1} - q_{n,0}\,K_{n,0}) - 2^5 q_{n,1}$$

$$\partial(\varepsilon_{n,2} - q_{n,1}(\varepsilon_{n,1} - q_{n,0}\,\varepsilon_{n,0}) - q^3_{n,0}\,\varepsilon_{n,0}) =$$

$$2^{\alpha(n)}(K_{n,2} - q_{n,1}K_{n,1} + (q_{n,1}q_{n,0} - q^3_{n,0})K_{n,0}) - 2^6 q_{n,2}$$

and by induction for $i \leq \alpha(n)-4$ we can find a polynomial

$$p_i = \varepsilon_{n,i} + \sum_{j=0}^{i-1} A_j(q_{n,0}, \cdots, q_{n,i-1})\,\varepsilon_{n,j}$$

with

$$\partial(p_i) = 2^{\alpha(n)}(K_{n,i} + \sum_{j=0}^{i-1} A_j(q_{n,0}, \cdots, q_{n,i-1})\,K_{n,j}) - 2^{4+i}q_{n,i}.$$

For $i > \alpha(n)-4$ we can find polynomials

$$q_i = \varepsilon_{n,i} + \sum_{j=0}^{\alpha(n)-4} B_j(q_{n,0}, \cdots, q_{n,\alpha(n)-4})\varepsilon_{n,j}$$

with

$$\partial(q_i) = 2^{\alpha(n)}\left[K_{n,i} + \sum B_j(q_{n,0} \cdots q_{n,\alpha(n)-4})K_{n,j} - 2P(q_{n,\alpha(n)-3}, \cdots, q_{n,i})\right].$$

Now an easy change of basis gives the proposition.

Finally, we point out that similar calculations can be given for the subspectral sequence 13.21 using 13.15 and we have

THEOREM 13.25. *The subspectral sequence* 13.21 *is isomorphic to* $\bigotimes_{n\ odd} E_*^*(A_n)$ *when the structure of* $E_*^S(A_n)$ *is as given in* 13.23 *and* 13.24.

CHAPTER 14
THE TYPES OF TORSION GENERATORS

In this chapter we summarize briefly the constructions of various types of torsion generators for Ω_*^{PL}. Those discussed in Section B coming from relations involving the Milnor manifolds seem intriguing. Next, we apply these constructions to obtain information about the unoriented bordism rings $\mathfrak{N}_*^{PD}$ and $\mathfrak{N}_*^{PL}$. Finally, in Section D we summarize some work of Ligaard, Mann, May and Milgram on the structure of odd torsion in $\Omega^{PL}{}_*(pt)$.

A. *Torsion generators, suspension, and the map* η

In 8.1, 8.2 we defined the map

$$\eta : \Omega_*(G/PL) \to \Omega_*^{PL}$$

which restricts to define

$$\mu : \Omega_*(G/O) \to \Omega_* .$$

We note the naturality property of μ which follows from 8.3

LEMMA 14.1. *The diagram below commutes*

$$\begin{array}{ccc} \Omega_*(G/O) & \xrightarrow{\ \mu\ } & \Omega_* \\ \Big\downarrow{\scriptstyle i_*} & & \Big\downarrow{\scriptstyle \tilde{j}} \\ \Omega_*(G/PL) & \xrightarrow{\ \eta\ } & \Omega_*^{PL} \end{array}$$

where i_* *is induced by the natural map of spaces and* $\tilde{j}$ *is the usual inclusion.*

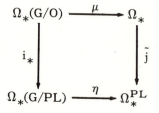

We have already pointed out in Chapter 8 that the only torsion in $\Omega_*(G/PL)$ is 2-torsion, and that η is a module map (where Ω_*^{PL} becomes an Ω_* module under $\tilde{j}$) and so also is μ. The next lemma computes μ on $\mathrm{Tor}\,\Omega_*(G/O)$.

LEMMA 14.2. *Let* $\{M, f\} \in \mathrm{Tor}\,\Omega_*(G/O)$ *then* $\{M\} - \mu\{M, f\} = 0$.

Proof. Since μ is a homomorphism $\mu\{M, f\}$ is torsion in Ω_*. If $\{M, f\}$ is odd torsion, then, since there is no odd torsion in Ω_* we have $\{M\} = 0$ and $\{\tilde{M}\} = 0$ in Ω_* so the theorem is true. If $\{M, f\}$ represents 2-torsion, we show the Stiefel-Whitney numbers of $M, \tilde{M}$ are equal. Indeed, $\nu(\tilde{M}) = \pi^*(\nu(M)) - \pi^* f^*(\gamma_{G/O})$. But $\gamma_{G/O} = j^*(\gamma)$ where $j: G/O \to BO$ is the usual map. However, in mod. 2 cohomology $j^* \equiv 0$ in positive degrees, hence $W(j^*(-\gamma)) = j^*(W(\gamma)) = 1$. Thus

$$W(\nu(\tilde{M})) = \pi^* W(\nu(M))$$

and since π has degree 1, the Stiefel-Whitney numbers of $M, \tilde{M}$ are equal. But the map

$$\mathrm{Tor}\,\Omega_* \to \Omega_* \to \mathfrak{N}_*,$$

where $\mathfrak{N}_*$ is the unoriented cobordism ring, is an injection [142] and it is well known that elements in $\mathfrak{N}_*$ are distinguished by their Stiefel-Whitney numbers (cf. Chapter 1.C).

COROLLARY 14.3. *If* $\{M, f\} \in \mathrm{Tor}\,\Omega_*(G/O)$ *and* $\{M\} = 0$ *in* Ω_* *then any representative* $\tilde{M}$ *for* $\mu\{M, f\}$ *is also a differentiable boundary.*

Now suppose $\alpha \in \mathrm{Tor}\,\Omega_*(G/O)$ and in 14.1 $i_*(\alpha) = \mu(\alpha) = 0$, then if (M, f) represents α and $\pi: \tilde{M} \to M$ is the associated degree 1 normal map, on the one hand $\tilde{M}$ is differentiably the boundary of some differentiable manifold W and on the other hand (M, f) is PL normally bordant to 0. That is, there is a differentiable manifold W' with $\partial W' = M$ and a map

$F: W' \to G/PL$ extending f, so W' is covered by a normal bordism $\tilde{W}'$ of $\tilde{M}$ to 0. (Here $\tilde{W}'$ is a PL manifold and probably not differentiable.)

DEFINITION 14.4. Let $a \in \mathrm{Tor}\,\Omega_*(G/O)$ satisfy $i_*(a) = \mu_*(a) = 0$, then the 'suspension' $\sigma(a)$ of a contained in $\Omega_*^{PL}(pt)$ is the bordism class of the PL manifold $\tilde{W}' \cup_{\tilde{M}} (-W)$. It is well defined as a coset of $\mathrm{im}(\tilde{j}) + \mathrm{im}(\eta)$.

LEMMA 14.5. *Let $a = \{M, f\}$ satisfy the conditions above, suppose also that $f_*([M])$ is an indecomposable in $H_*(G/O, Z/p)$ and in the Serre spectral sequence of the fibering*

14.6 $$G/O \longrightarrow BO \times G/PL \xrightarrow{\pi} BPL$$

there is an element $b \in H_*(BPL, Z/p))$ with $d(b) = f_*([M])$ then

$$\nu_*([\sigma(a)]) = b + \gamma$$

where γ is in $\mathrm{im}(\pi_*)$.

Proof. The PL-normal bundle of $\tilde{W}' \cup_{\tilde{M}} (-W)$ is $\pi^*(\nu_{W'} - F^*(\zeta)) \cup \nu_W$ where $\pi: \tilde{W}' \to W'$ is the normal map associated with the extension $F: W' \to G/PL$ of f. Thus we have a commutative diagram

$$\begin{array}{ccc} \tilde{W}' \cup_{\tilde{M}} (-W) & \xrightarrow{\nu} & BPL \\ \downarrow & & \downarrow \\ \tilde{W}'/\tilde{M} & \xrightarrow{F} & B(PL/O) . \end{array}$$

The rest is now formal from the definitions.

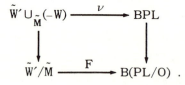

REMARK 14.7. So far we have considered torsion arising from $\eta(\mathrm{Tor}\,(\Omega_*(G/PL)))$, which is only 2-torsion since $\mathrm{Tor}\,\Omega_*(G/PL)$ is only 2-torsion and torsion coming from σ. Modulo its indeterminacy $\sigma(a)$ is also torsion (since the indeterminacy contains the torsion free parts), but

here we obtain odd primary torsion as well and 14.5 shows that a large number of odd torsion generators can be constructed in this way.

The third type of element arises in case $\alpha \in \Omega_*(G/O)$ is odd dimensional but $i_*(\alpha) = \beta \neq 0$ in $\Omega_*(G/PL)$. If $\beta^2 = 0$ in $\Omega_*(G/PL)$ we obtain the diagram of degree 1 normal maps

14.8
$$\begin{array}{ccc} \tilde{M}\times\tilde{M} & \subset & \tilde{W} \\ \downarrow & & \downarrow \\ M\times M & \subset & W \end{array}$$

where $M\times M = \partial W$, $\tilde{M}\times\tilde{M} = \partial\tilde{W}$. Of course, for G/TOP the same diagram 14.8 holds in dimensions ≥ 5 as well. As in 14.3 we can assume M is differentiably a boundary, $\tilde{M} = \partial\tilde{N}$ so we can construct the oriented manifold

14.9
$$\tilde{M}\times\tilde{N} \cup_{\tilde{M}\times\tilde{M}} (-\tilde{W})$$

and the bordism class of this manifold is well defined up to an element in $\mathrm{im}(\eta)$.

REMARK 14.10. The manifold construction 14.9 corresponds to the homology construction $h_{i,j}$(i odd) in 13.11, and it is not hard to see that the Hurewicz images correspond as well.

There are two other types of generators in the 2 primary part of $\Omega_*^{PL}(pt)$ obtained by considering the torsion part of the diagram in 14.1. They correspond to the classes J_{ij}, L_{ij} in 13.12. Since the chain descriptions of these classes are quite involved, we don't attempt the construction here.

B. *Torsion coming from relations involving the Milnor manifolds*

The remaining torsion generators which we are about to discuss occur only at the prime 2 and occur because in dimensions of the form 4n for which $\alpha(n) - 4 < \nu_2(n)$ the Milnor manifold M^{4n} and its divided powers can rationally be expressed as polynomials in the differentiable torsion free generators, and decomposable expressions involving lower dimensional M^{4n}'s and their divided powers (cf. Chapter 8.B, and Chapter 10.C).

What this means is that for $\alpha(n)-4<\nu_2(n)$ there is a polynomial P in torsion free generators so that $M^{4n}-P$ represents torsion in the PL bordism ring. We use the results in Chapter 13 in particular 13.24 to determine the order of $M-P$ by looking at its Hurewicz image in $H_*(BTOP, Z/2)$ which is clearly the image of K_{4n}, and the torsion order of this element is $2^{\alpha(n)}$ so we have

THEOREM 14.11. *There is a polynomial of the form* $(\varepsilon M^{4n}-P_n)$ *with* ε *odd and* P_n *involving only the torsion free generators so that modulo decomposable torsion the order of* $(\varepsilon M^{4n}-P_n)$ *is* $2^{\alpha(n)}$ *in* $\Omega_*^{TOP}(pt)$ *for* $\alpha(n)-4<\nu_2(n)$.

Moreover, in $\Omega_*^{PL}(pt)$ *the same expression has order*

$$2^{\alpha(n)} \quad \text{for} \quad \alpha(n)>1$$

$$4 \quad \text{for} \quad \alpha(n)=1\,.$$

(The first statement follows from 13.24. The second from the fibering $BPL \to BTOP \to K(Z/2,4)$ when we calculate the resulting Serre spectral sequence and compare it to the Bockstein spectral sequence.)

Here are two specific examples to illustrate 14.11. Rationally, $7M^8$ has the same Pontrjagin numbers as $200(CP^2\times CP^2)-144\, CP^4 = 8(25(CP^2\times CP^2)-18\, CP^4)$. (Notice that the Hurewicz image of $(25\, CP^2\times CP^2-18\, CP^4)$ is the smallest multiple of the integral primitive s_8 in $H_8(BO, Z)/\mathrm{Tor}$ which is represented by the normal map of a differentiable manifold, then check indexes).

Also, since the group of homotopy 7-spheres $\Gamma_7 = Z/28$, $28M^8$ is differentiable, so

14.12 $$4(7M^8-200(CP^2\times CP^2)+144\, CP^4)$$

is cobordant to 0 in Ω_*^{PL}. In fact, from [143] we know this is best possible. However, in Ω_8^{TOP} we have

14.13 $$2(7M^8 - 200(CP^2 \times CP^2) + 144\ CP^4) \sim 0\ .$$

In dimension 12 the torsion class is found by the same method and has order 4 in both Ω_*^{PL} and Ω_*^{TOP}. Its explicit expression is $31M^{12} - 1620\ CP^6 + 5292(CP^4 \times CP^2) + 3920(CP^2)^3$.

C. *Application to the structure of the unoriented bordism ring* $\mathfrak{N}_*^{PD}, \mathfrak{N}_*^{PL}$

The "suspension" of singular manifolds used above also gives information on the geometric structure of the unoriented cobordism theories $\mathfrak{N}_*^{PD}$ and $\mathfrak{N}_*^{PL}$ of Poincaré duality spaces and PL-manifolds.

First, recall that geometrically $\pi_*(MG)$ is the cobordism theory of normal spaces (Quinn [113]) and from [29] that

14.14 $$\pi_*(MG) = \mathfrak{N}_*^{Diff} \otimes H_*(B(G/O); Z/2)\ .$$

Let $f: M^n \to SG$ be a smooth singular manifold in SG and let $\pi: \tilde{M}^n \to M^n$ denote its associated surgery problem. Suppose also that M^n and hence $\tilde{M}^n$ are smooth boundaries, $M = \partial W$, $\tilde{M} = \partial \tilde{W}$. Then $\tilde{W} \cup_\pi W$ is a normal space and its normal fibration

$$\nu: \tilde{W} \cup_\pi W \to BG$$

represents the suspension of $f_*([M^n])$ at least in the quotient $H_*(B(G/O); Z/2) = H_*(BG; Z/2) // H_*(BO; Z/2)$. But this ring is generated by the image under the homology suspension of $H_*(G/O; Z/2)$. Thus $\pi_*(MG)$ is generated by the normal spaces $\{\tilde{W} \cup_\pi W\}$.

When $\pi: \tilde{M}^n \to M^n$ above is cobordant to a homotopy equivalence then the class of $\{\tilde{W} \cup_\pi W\}$ of course represents an element of $\mathfrak{N}_*^{PD} \subset \pi_*(MG)$. Let

$$s_K : \mathfrak{N}_*^{Diff}(SG) \to Z/2$$

be the Kervaire invariant. Every element of $H_*(SG; Z/2)$ of degree $\neq 2$

is represented by an element in the kernel of s_K and in dimensions at least 5 the associated suspension $\{\tilde{W} \cup_\pi W\}$ belongs to $\mathfrak{N}_*^{PD}$. The elements $\{\tilde{W} \cup_\pi W\}$ with $\pi: \partial\tilde{W} \to \partial W$ a homotopy equivalence thus generate a large part of $\mathfrak{N}_*^{PD}$ but not all as e.g. the product of such an element with the 3-dimensional normal space $D^2 \times S^1 \cup_\pi D^3$, where $\pi: S^1 \times S^1 \to S^2$ is the non-trivial surgery problem, is not necessarily of this form.

We next turn to the PL-case. We have

14.15 $$\mathfrak{N}_*^{PL} = \mathfrak{N}_*^{Diff} \otimes H_*(B(PL/O); \mathbf{Z}/2)$$

but this time the suspension

$$\sigma: H_*(PL/O; \mathbf{Z}/2) \to QH_*(B(PL/O); \mathbf{Z}/2)$$

is not surjective. However, the remaining generators are in the image of the natural map

$$j_*: H_*(G/PL; \mathbf{Z}/2) \to H_*(B(PL/O); \mathbf{Z}/2)\ .$$

Using smoothing theory instead of surgery theory one sees that elements of $\mathrm{Im}(\sigma)$ are represented by composites

$$\tilde{W} \cup_\pi W \xrightarrow{\nu} BPL \longrightarrow B(PL/O)$$

where $\tilde{W}$ and W are smooth manifolds, $\pi: \partial\tilde{W} \to \partial W$ a PL-homeomorphism and ν denotes the PL-normal bundle of $\tilde{W} \cup_\pi W$.

Every element of $H_*(G/PL; \mathbf{Z}/2)$ (of degree at least 5) can be represented by a smooth singular manifold $f: M^n \to G/PL$ whose associated surgery problem $\pi: \tilde{M}^n \to M^n$ is a homotopy equivalence. In fact, the non-spherical classes can be represented by products of real projective spaces. But

$$\tilde{M}^n \xrightarrow{\nu} BPL \longrightarrow B(PL/O)$$

is exactly $j_*(f_*[M^n])$ and we therefore have

THEOREM 14.16. $\mathfrak{N}_*^{PL}$ *is multiplicatively generated by manifolds of the following two types*

a) *Homotopy triangulations of products of real Projective spaces.*

b) $\tilde{W} \cup_\pi W$, *with* $\tilde{W}$ *and* W *smooth and* $\pi : \partial\tilde{W} \to \partial W$ *a PL-homeomorphism.*

D. p-*torsion in* Ω_*^{PL} *for* p *odd*

In recent work [158], [159], [160], much progress has been made in studying the structure of the odd torsion in the PL-bordism rings. Work in this area was initiated primarily by F. Peterson [110].

The basic step is to obtain the structure of $H^*(MSPL, Z/p)$ as a module over the Steenrod algebra $\mathcal{A}_p$. In order to do this, note the formula,

$$\mathcal{P}^i(a \cup U) = \sum \pm \mathcal{P}^j(a) \cup \mathcal{P}^{i-j}(U) .$$

This implies, by Sullivan's splitting results 5.12, and 5.20; (since we can assume the structure of $H^*(B\,\mathrm{cok}\,J, Z/p)$ over $\mathcal{A}_p$ by [86], [139]) that it suffices to know the action of $\mathcal{A}_p$ on U in $H^*(M\,\mathrm{cok}\,J, Z/p)$.

Preliminary work on this problem was carried out in [140] where Tsuchiya proved a conjecture of F. Peterson.

THEOREM 14.17 (Tsuchiya). *Let* $Q = \beta\mathcal{P}_1 - \mathcal{P}_1\beta$, *and* $Q_0 = \beta$ *in* $\mathcal{A}_p$ *then the map*

$$\alpha : \mathcal{A}_p \to H^*(MSPL, Z/p)$$

defined by $\alpha(\mathcal{P}^I) = \mathcal{P}^I U$ *has kernel precisely the left ideal generated by* Q_0, Q_1.

Unfortunately, this is not quite sharp enough since we need to actually know precisely which elements occur as $\mathcal{P}^I(U)$, not merely that they are non-zero.

To sharpen this result the first step is to consider the map

$$MSPL \to MSG$$

and to evaluate the action of $\mathcal{A}_p$ in $H^*(MSG, Z/p)$ as far as possible.

Next we consider the map

$$M(\Sigma G) \to MSG$$

where the bundle on ΣG is induced from the universal bundle on BG by the usual inclusion $\sigma: \Sigma G \to BG$. The map $\lambda: (\mathcal{A}_p)_i \to H^{i-1}(G, Z/p)$ defined by $\lambda(\mathcal{P}^I) = \sigma^*[(\mathcal{P}^I U) \cap U]$ is now analyzed, and its image is shown to lie in the set of elements primitive under both loop sum and composition. This set is very small and effective calculations can be carried out. Finally, the desired information on α is obtained by using the Hopf algebra structure of $H^*(BG, Z/p)$, and the relation of algebra generators to $H^*(G, Z/p)$ by σ^*.

The main technical result of [158], [159] is

THEOREM 14.18. *There exists a sub-Hopf algebra*

$$B = E\{V_2, W_2, W_3, \cdots\} \otimes \Gamma_p\{Q_0(V_2), Q_1(V_2), \underline{W}_3, \cdots\}$$

in $H^*(B\operatorname{cok} J, Z/p)$ *so that*

(i) $Q_0(Q_1(V_2) = W_2$

(ii) $Q_0(\underline{W}_j) = Q_1(\gamma \underline{W}_{j-1}) = Q_2(\gamma^2(\underline{W}_{j-2})) = \cdots = Q_{j-1}(\gamma^{j-2}Q_0(V_2)) = W_j,\ j \geq 3$

(iii) $Q_1(\gamma^r(Q_0(V_2))) = (\gamma^{r-1}(Q_0(V_2)))^{p-1} \cdots (\gamma(Q_0(V_2)))^{p-1} \cdot Q_0(V_2)^{p-1} \cdot (Q_1Q_0(V_2))$

(iv) $Q_0(\gamma^r(\underline{W}_j)) = (\gamma^{r-1}(\underline{W}_j))^{p-1} \cdots (\gamma(\underline{W}_j))^{p-1}(\underline{W}_j)^{p-1}W_j,\ j \geq 2$ *and* $\underline{W}_2 = Q_1V_2$

(v) $Q_1(\gamma^r(\underline{W}_j)) = 0 \quad r \geq j$

(vi) $Q_0(\gamma^r(Q_0(V_2))) = 0 \quad r \geq 0$

and in $H^*(M\operatorname{cok} J, Z/p)$, $Q_j(U) = W_j \cup U \quad j \geq 2$.

Note that inductively $Q_i = Q_{i-1}\mathcal{P}^{p^{i-1}} - \mathcal{P}^{p^{i-1}}Q_{i-1}$, and $\Gamma_p\{x_1, x_2, \cdots\}$ is the divided power algebra over Z/p (that is, its dual is a primitively generated polynomial algebra).

This theorem is sufficient to obtain the E^2 term of the Adams spectral sequence converging to $\pi_*(MSPL) \otimes \hat{Z}_p$. Unfortunately, there are many differentials, and vastly complex systems of generators. There also appears to be p-torsion of all orders. However, in low dimensions calculations can be made.

As an example, the 3-torsion in Ω_*^{PL} is given through dimension 50 by the table

14.19

Dimension	3-Torsion in Ω_*^{PL}
11	$\mathbf{Z}/3$
23	$(\mathbf{Z}/3)^2$
27	$\mathbf{Z}/3 + \mathbf{Z}/9$
34	$\mathbf{Z}/3$
35	$(\mathbf{Z}/3)^2 + \mathbf{Z}/9$
38	$\mathbf{Z}/3$
39	$(\mathbf{Z}/3)^3 + \mathbf{Z}/27$
43	$(\mathbf{Z}/3)^2 + \mathbf{Z}/81$
46	$(\mathbf{Z}/3)^2$
47	$(\mathbf{Z}/3)^6$
50	$\mathbf{Z}/3$.

APPENDIX

THE PROOFS OF 13.12, 13.13 AND 13.15

We begin by constructing some universal models.

DEFINITION A.1. Let B_{2s+1} be the fiber of the map

$$B_{2s+1} \xrightarrow{j} K(Z/2, 2s+1) \xrightarrow{2Sq^{2s}\iota} K(Z/4, 4s+1) .$$

Note that $(2Sq^{2s}\iota)^*(\beta_2 \iota_{4s+1}) = Sq^{2s+1}\iota$ since $Sq^1(Sq^{2s}\iota) = Sq^{2s+1}\iota = \iota^2$ so $\delta(Sq^{2s}\iota) = 2(\iota^2)$ on the cochain level and $\delta\, 2(Sq^{2s}\iota) = 4(\iota^2)$, but $\beta_2 \iota_{4s+1}$ goes to $\frac{1}{4}\delta(2\, Sq^{2s}\iota)$.

Now consider the fibration

A.2 $$K(Z/4, 4s) \xrightarrow{\hat{j}} B_{2s+1} \xrightarrow{j} K(Z/2, 2s+1) .$$

Let γ be the fundamental class on the fiber. Then γ is an infinite cycle modulo 2 and we have

LEMMA A.3. *There is a cohomology class* $\gamma' \in H^{4s}(B_{2s+1}, Z/2)$ *so* $\hat{j}^*(\gamma') = \gamma$ *and* $Sq^1(\gamma') = j^*(Sq^{2s}\iota_{2s+1})$.

Proof. There is a cocycle representative for γ' with $\delta(\gamma') = j^*(2(Sq^{2s}\iota)) + 4V$ where $\hat{j}^*(V)$ represents $\beta_2\gamma$. But $2(Sq^{2s}\iota) + 4V = 2((Sq^{2s}\iota) + 2V)$.

COROLLARY A.4. *In* $H^{2s+1}(B_{2s+1}; Z/2)$ *the fundamental class* $j^*(\iota)$ *satisfies* $(j^*\iota)^2 = 0$. *In particular* $(j^*(\alpha))^2 = 0$ *for all* $\alpha \in H^*(K(Z/2, 2s+1); Z/2)$.

This follows since $(Sq^I\iota)^2 = Sq^{2I}(\iota^2)$ in $K(Z/2, 2s+1)$, and with mod. 2 coefficients $x \to x^2$ is a homomorphism. Also, since $Sq^1 Sq^{2s}\iota = \iota^2$ and $Sq^{2s}j^*(\iota) = Sq^1(\gamma')$ we obtain the first statement.

Next, we construct a space which is considerably better known (see e.g. [1], [93]).

DEFINITION A.5. E_{2s} is the fiber in the map

$$E_{2s} \xrightarrow{j} K(Z/2, 2s) \xrightarrow{\iota^2} K(Z/2, 4s).$$

It has the property that in the fibering

A.6 $$K(Z/2, 4s-1) \xrightarrow{\hat{j}} E_{2s} \xrightarrow{j} K(Z/2, 2s)$$

there is a class $g \in H^{4s}(E_{2s}; Z/2)$ with $\hat{j}^*(g) = Sq^1(\iota_{4s-1})$ and

A.7 $$Sq^1(g) = j^*(\iota\, Sq^1\iota + Sq^{2s}\, Sq^1\iota),\ \bar{\psi}(g) = j^*(\iota) \otimes j^*(\iota).$$

Now consider the fibering

A.8 $$H_{2s+1} \longrightarrow B_{2s+1} \xrightarrow{\alpha} K(Z/2, 2s+1) \times K(Z/2, 4s)$$

with K-invariants ι, γ'. In homotopy α_* is an isomorphism in dimension $2s+1$, and in dimension $4s, \alpha_*: Z/4 \to Z/2$ is onto, hence H_{2s+1} is the Eilenberg-MacLane space $K(Z/2, 4s)$. On the other hand, the cohomology Leray-Serre spectral sequence for the fibering

A.9 $$K(Z/2, 2s) \times K(Z/2, 4s-1) \to H_{2s+1} \to B_{2s+1}$$

is highly non-trivial. Let $\theta_{2s}, \theta_{4s-1}$ be the fundamental classes on the fiber. Then from A.3

$$Sq^1\,\theta_{4s-1} + Sq^{2s}\,\theta_{2s} \quad \text{and} \quad \iota_{2s+1} \otimes \theta_{2s}$$

are the two lowest dimensional surviving cycles. (Note that $\iota_{2s+1} \otimes \theta_{2s}$

occurs in the middle of the grid and $Sq^1\theta_{4s-1} + Sq^{2s}\theta_{2s}$ occurs on the vertical edge corresponding to the fiber.) Since $H_{2s+1} = K(\mathbf{Z}/2, 4s)$ it follows that

$$Sq^1\{Sq^1\theta_{4s-1} + Sq^{2s}\theta_{2s}\} = \{\iota_{2s+1} \otimes \theta_{2s}\}.$$

LEMMA A.10. *Let* $F \to E \to B$ *be any fibering. Suppose* $x \in H^{2s+1}(B; \mathbf{Z}/2)$ *is in the transgressive image,* $(x = d_{2s+1}(y))$ *in the Leary-Serre spectral sequence for* E. *Suppose also that*

$$Sq^{2s}x = Sq^1\gamma$$

with $\gamma = d_{4s}(z)$, *then* $Sq^1z + Sq^{2s}y$, $x \otimes y$ *are infinite cycles and represent classes* A, B *in* $H^*(E; \mathbf{Z}/2)$ *with* $Sq^1(A) = B$.

Proof. The situation above in A.9 is universal for these properties.

We now apply a similar procedure using E_{2s}. Consider the fibering

A.11 $$F_{2s} \to E_{2s} \to K(\mathbf{Z}/2, 2s)$$

with K-invariant ι_{2s}. Clearly, F_{2s} is $K(\mathbf{Z}/2, 4s-1)$. On the other hand, continuing A.11 to the left we have the fibration

A.12 $$K(\mathbf{Z}/2, 2s-1) \to F_{2s} \to E_{2s}$$

with Leray-Serre spectral sequence having as its lowest degree survivors $\iota_{2s} \otimes \theta_{2s-1}$ and g (where g is discussed after A.6). Also θ_{2s-1} is the fundamental class on the fiber. Thus $Sq^1\{\iota_{2s} \otimes \theta_{2s-1}\} = g$ and we have

LEMMA A.13. *Let* $F \to E \to B$ *be any fibering. Suppose* $x \in H^{2s}(B; \mathbf{Z}/2)$ *is* $d_{2s}(y)$ *in the Leray-Serre spectral sequence. Also, suppose* $x^2 = 0$ *so*

$$x\, Sq^1x + Sq^{2s}Sq^1x = Sq^1(g)$$

in $H^*(B)$. *Then* $x \otimes y$ *and* g *are infinite cycles and* $Sq^1\{x \otimes y\} = g$.

The proof of 13.12

The classes $y_{i,j}$ and f_{2i+1} in 12.4 are dual to primitive cohomology classes $Y_{i,j}$ and $\phi_{2i+1}(=Y_{0,j})$, and $Y_{i,j}^2 = 0$ for all $0 \leq i \leq j$ since $H_*(B(G/O); Z/2)$ is primitively generated. In particular when i is odd $\deg(Y_{i,j})$ is even and we may apply A.13 to the fibering

$$G/TOP \times BO \to BTOP \to B(G/O)$$

and the class $Y_{i,j}$ and we obtain the first statement of 13.12 (á) and the second statement of 13.12 (b) on using the formula

$$x\, Sq^1 x + Sq^{2s}\, Sq^1 x = Sq^1 g$$

in A.13 to show that $y_{i,j}^2$ is dual to g and reversing arrows on going from cohomology to homology.

Similarly, if i is even, then

$$Sq^{i+2j}(Y_{i,j}) = Y_{2i,2j} = Sq^1 Y_{2i-1,2j}$$

and the conditions for applying A.10 are satisfied. This gives the second part of 13.12 (a) and the first part of 13.12 (b).

We need further models before proving 13.13 and 13.15.

DEFINITION A.14. Let M_{2s+1} be the fiber in the fibration

$$M_{2s+1} \xrightarrow{\pi} K(Z/2, 2s+1) \xrightarrow{\beta\, Sq^{2s}(\iota)} K(Z, 4s+2)\,.$$

Then M_{2s+1} is universal for the property that $Sq^{2s}(\iota)$ be the reduction of an integral class. We let $\{Sq^{2s}(\iota)\}$ denote the universal integral class in $H^{4s+1}(M_{2s+1}; Z)$. It restricts to twice the fundamental class on the fiber in

$$K(Z, 4s+1) \to M_{2s+1} \to K(Z/2, 2s+1)$$

as one sees directly from the diagram

A.15
$$\begin{array}{ccccccc}
K(Z,4s+1) & \longrightarrow & M_{2s+1} & \longrightarrow & K(Z/2,2s+1) & \longrightarrow & K(Z,4s+2) \\
\downarrow{\scriptstyle Id} & & \downarrow{\scriptstyle \{Sq^{2s}\iota\}} & & \downarrow{\scriptstyle Sq^{2s}\iota} & & \downarrow{\scriptstyle Id} \\
K(Z,4s+1) & \xrightarrow{2} & K(Z,4s+1) & \longrightarrow & K(Z/2,4s+1) & \xrightarrow{\beta} & K(Z,4s+2)
\end{array}$$

Dually, the Hurewicz image z of the generator of $\pi_{4s+1}(M_{2s+1})$ is divisible by 2 and $\frac{1}{2}z$ is dual to $\{Sq^{2s}\iota\}$. Consider the fibration

A.16
$$F_{2s+1} \longrightarrow M_{2s+1} \xrightarrow{\phi_s} K(Z/2, 2s+1) \times K(Z \oplus Z, 4s+1)$$

where the first projection of ϕ_s is the map π from A.14 and

A.17
$$\begin{aligned} \phi_s^*(I_1) &= 2^{\alpha(s)-1}\{Sq^{2s}(\iota)\} \\ \phi_s^*(I_2) &= 2^{\nu_2(s)+3}\{Sq^{2s}(\iota)\}\,. \end{aligned}$$

Here I_1, I_2 are the fundamental classes of $K(Z \oplus Z, 4s+1) = K(Z, 4s+1) \times K(Z, 4s+1)$.

LEMMA A.18. $F_{2s+1} = K(Z \oplus Z/2^{\lambda(s)}, 4s)$ *where* $\lambda(s) = \min(\alpha(s), 4+\nu_2(s))$.

Proof. We check in homotopy using A.15: $(\phi_s)_*$ is an isomorphism in dimension $2s+1$ while in dimension $4s+1$

$$(\phi_s)_*: Z \to Z \oplus Z$$

is given by $(\phi_s)_*(e) = 2^{\alpha(s)}g_1 \oplus 2^{\nu_2(s)+4} g_2$ and so $\pi_i(F_{2s+1}) = 0$, $i \neq 4s$ while $\pi_{4s}(F_{2s+1}) \cong \operatorname{coker}(\phi_s)_* = Z \oplus Z/2^{\lambda(s)}$ and A.18 follows.

Now, consider the Serre spectral sequence for the fibering

A.19
$$K(Z/2, 2s) \times K(Z \oplus Z, 4s) \to F_{2s+1} \to M_{2s+1}\,.$$

Let J_1, J_2 be the 4s dimensional fundamental classes of the fiber (with $Z/2$ coefficients).

LEMMA A.20. *If* $\alpha(s) > 1$, *then in the Serre spectral sequence for* A.19 I_1, I_2 *are infinite cycles on the fiber as is* $\iota_{2s+1} \otimes \iota_{2s}$ *in the middle of the grid. Moreover* $\iota_{2s+1} \otimes \iota_{2s}$ *represents the image of* $\beta_s(J_1)$ *if* $\alpha(s) \le 4+\nu_2(s)$, *or* $\beta_s(J_2)$ *if* $4+\nu_2(s) \le \alpha(s)$, *where* β_s *in each case is the* $2^{\lambda(s)}$ *Bockstein.*

Proof. Certainly $\iota_{2s+1} \otimes \iota_{2s}$ is an infinite cycle and we easily check that E_∞ has the form

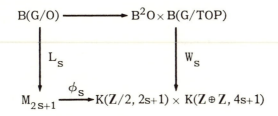

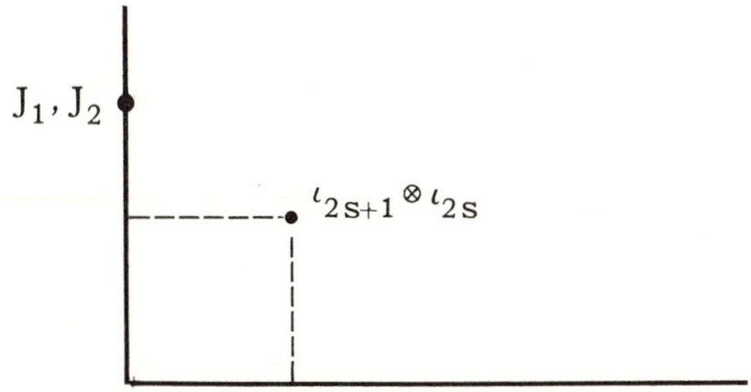

in total dimensions $\le 4s+1$ with all other positions zero. Hence $\iota_{2s+1} \otimes \iota_{2s}$ must be the mod.2 reduction of the $2^{\lambda(s)}$ Bockstein and A.20 follows.

LEMMA A.21. *For each* s *there are maps*

$$W_s : B^2O \times B(G/TOP) \to K(\mathbf{Z}/2, 2s+1) \times K(\mathbf{Z} \oplus \mathbf{Z}, 4s+1)$$

$$L_s : B(G/O) \to M_{2s+1}$$

so that the diagram

$$\begin{array}{ccc} B(G/O) & \longrightarrow & B^2O \times B(G/TOP) \\ \downarrow L_s & & \downarrow W_s \\ M_{2s+1} & \xrightarrow{\phi_s} & K(\mathbf{Z}/2, 2s+1) \times K(\mathbf{Z} \oplus \mathbf{Z}, 4s+1) \end{array}$$

commutes. Moreover, we may assume $L_s^*(\iota_{2s+1}) = \phi_{2s+1}$ *and* $W_s^*(I_1) = k_{4s+1}$, $W_s^*(I_2) = h_{4s+1}$, *in the notation of Chapter* 12.6.

Proof. From 12.5 it follows that $Sq^{2s}\phi_{2s+1} = \phi_{4s+1}$ and ϕ_{4s+1} is the reduction of the integral class $\hat{\phi}_{4s+1}$ uniquely characterized in 12.10. The classes ϕ_{2^j-1} are in the image of $(Bs)^*$ and 12.6 then shows that the same is true for all the ϕ_{2s+1}. Now, A.21 follows from A.17, 12.14(i) and 12.15 once we note that $\nu_2(3^{2s}-1) = \nu_2(s)+3$.

The proofs of 13.13 *and* 13.15

The proof of 13.15 is direct from A.20 and A.21 as we can take the space Y_{4s} in 13.15 to be the fiber F_{2s+1} of $\phi_s : M_{2s+1} \to K(\mathbf{Z}/2, 2s+1) \times K(\mathbf{Z} \oplus \mathbf{Z}, 4s+1)$ and consider the map $\lambda_s : BTOP \to F_{2s+1}$ induced from the diagram in A.21.

It remains to prove $\partial_1(h_{0,j}) = 0$ for $j \neq 2^s$. First, note that $h_{0,j}$ is primitive, hence $\partial_1(h_{0,j})$ is primitive and for dimensional reasons must project non-trivially to $H_*(B(G/O); \mathbf{Z}/2)$ or be in the image of $H_*(BO \times G/TOP; \mathbf{Z}/2)$. If $\partial_1(h_{0,j})$ projects to $H_*(B(G/O); \mathbf{Z}/2)$ then so does $h_{0,j}$ which contradicts the fact that it is represented by an interior grid point. Now, if $\partial_1(h_{0,j}) \neq 0$ and belongs to $\mathrm{Im} H_*(BO \times G/TOP; \mathbf{Z}/2)$ then either $\partial_1(h_{0,j}) = \partial_1(a)$ with $a \in \mathrm{Im} H_*(BO \times G/TOP)$ and we change $h_{0,j}$ to $h_{0,j} + a$ or $\partial_1(h_{0,j}) = \partial_1(\lambda)$ for some λ at the same grid point as $h_{0,j}$. Using the models A.9 and A.12 above and the maps constructed in A.10, A.13 $h_{0,j}$ goes to zero under all of them so $\partial_1 h_{0,j} = 0$ or $\partial_1 h_{0,j} = \varepsilon K_{4j} + \tau h_{4j}$. But the map

$$\phi : BTOP \to F_{2s+1}$$

constructed in A.21 has $\phi_* h_{0,j}$, $\phi_*(K_{4j})$ and $\phi_*(a_{4j})$ all non-zero and $\partial_1 \phi_*(h_{0,j}) = 0$. Thus $\partial_1 h_{0,j} = 0$ and 13.13 follows.

BIBLIOGRAPHY

[1] J. F. Adams, "On the non-existence of elements of Hopf invariant one," Ann. of Math. 72 (1960), 20-104.

[2] ________, "On the groups J(X) I-IV," Topology 2 (1963), 181-195; Topology 3 (1964), 137-171, 193-222; Topology 5 (1966), 21-71.

[3] ________, "*Lectures on generalized cohomology*," Lecture Notes in Mathematics, vol. 99 Springer-Verlag (1969).

[4] ________, "Stable homotopy and generalized homology," Mathematics Lecture Notes, University of Chicago, Chicago (1970).

[5] ________, "*Lectures on Lie groups*," W. A. Benjamin (1969).

[6] J. Adem, "The relations of Steenrod powers on cohomology classes," *Symposium in honor of S. Lefschetz*, Princeton University Press, Princeton (1957), 191-238.

[7] S. Araki and T. Kudo, "Topology of H_n-spaces and H_n-squaring operations," Mem. Fac. Sci. Kyusyn Univ. (1956), 85-120.

[8] M. F. Atiyah, "Thom complexes," Proc. London Math. Soc., 11 (1961), 291-310.

[9] ________, "K-*theory*," W. A. Benjamin, (1968).

[10] M. F. Atiyah, R. Bott and A. Shapiro, "Clifford modules," Topology 3 (1964), suppl. 1, 3-38.

[11] M. F. Atiyah, F. Hirzebruch, "Cohomologie-Operationen und Characteristic Klassen," Math. Z. 77 (1961), 149-187.

[12] M. F. Atiyah and D. O. Tall, "Group representations, λ-rings and the J-homomorphism," Topology 8 (1969), 253-297.

[13] M. F. Atiyah and G. B. Segal, "Equivariant K-theory and completion," J. Diff. Geom 3 (1969), 1-18.

[14] M. F. Atiyah and G. B. Segal, "Exponential isomorphisms for λ-rings," Quart. J. Math. (1971), 371-378.

[15] M. G. Barratt, "A note on the cohomology of semigroups," J. Lond. Math. Soc. 36 (1961), 496-498.

[16] M. G. Barratt and S. Priddy, "On the homology of non-connected monoids and their associated groups," Comm. Math. Helv. 47 (1972), 1-14.

[17] M. G. Barratt and M. Mahowald, "The Arf invariant in dimension 62," (to appear).

[18] J. Becker and D. Gottlieb, "The transfer map and fiber bundles," Topology 14 (1975), 1-13.

[19] J. M. Boardman and R. Vogt, "*Homotopy invariant algebraic structures*," Lecture Notes in Mathematics 347 Springer-Verlag (1973).

[20] A. Borel, "Sur la cohomologie des espaces fibres principaux et des espaces homogénes de groupes de Lie compacts," Ann. of Math. 57 (1953), 115-207.

[21] Z. I. Borevich and I. R. Shafarevich, "*Number theory*," Academic Press (1966).

[22] E. Brieskorn, "Beispiele zur Differentialtopologie von Singularitäten," Invent. Math. 2 (1966), 1-14.

[23] W. Browder, "Torsion in H-spaces," Ann. of Math. 74 (1961), 24-51.

[24] ———, "Higher torsion in H-spaces," Trans. A.M.S. 108 (1963), 353-375.

[25] ———, "On differential Hopf algebras," Trans. A.M.S. 107 (1963), 153-176.

[26] ———, "The Kervaire invariant of a framed manifold and its generalization," Ann. of Math. 90 (1969), 157-186.

[27] ———, "*Surgery on simply connected manifolds*," Ergebnisse der Math. 65 Springer-Verlag (1972).

[28] ———, "Homology operations and loop spaces," Illinois J. Math. 4 (1960), 347-357.

[29] W. Browder, A. Liulevicius and F. Peterson, "Cobordism theories," Ann. of Math. 84 (1966), 91-101.

[30] E. H. Brown, "Cohomology theories," Ann. of Math. 75 (1962), 467-484.

[31] G. Brumfiel, "On the homotopy groups of BPL and PL/O," Ann. of Math. 88 (1968), 291-311.

[32] ————, "On the homotopy groups of BPL and PL/O, II, III," Topology 8 (1969), 305-311, Mich. J. Math. 17 (1970), 217-224.

[33] ————, "On integral PL-characteristic classes," Topology 8 (1969), 39-46.

[34] G. Brumfiel and I. Madsen, "Evaluation of the transfer and the universal surgery classes," Invent. Math. 32 (1976), 133-169.

[35] G. Brumfiel, I. Madsen and R. J. Milgram, "PL characteristic classes and cobordism," Ann. of Math. 97 (1972), 82-159.

[36] G. Brumfiel and J. Morgan, "Homotopy theoretic consequences of N. Levitts obstruction theory to transversality for spherical fibrations," Pacific J. Math. (1976), 1-100.

[37] H. Cardenas, "El algebra de cohomologia del grupo simetrico de grado p^2," Bol. soc. Mat. Mex. 10 (1965), 1-30.

[38] H. Cartan, "Sur les groupes d'Eilenberg-MacLane $H(\pi,n)$, I, II," Proc. Nat. Acad. Sci., U.S.A. 40 (1954), 467-471, 704-707.

[39] J. Cerf, "*Sur les diffeomorphismes de la sphere de dimension trois* $(\Gamma_4 = 0)$," Lecture Notes in Mathematics 53, Springer-Verlag (1968).

[40] P. E. Conner and E. E. Floyd, "*Differentiable Periodic Maps*," Springer-Verlag (1964).

[41] ————, "Fibering within a cobordism class," Michigan J. Math. 12 (1965), 33-47.

[42] ————, "*The relation of cobordism to* K-*theories*," Lecture Notes in Mathematics 28, Springer-Verlag (1966).

[43] B. Conrad, "Extending free circle actions on spheres to S^3 actions," Proc. A.M.S. 27 (1971), 168-174.

[44] L. E. Dickson, "A fundamental system of invariants of the general modular linear group with a solution of the form problem," Trans. A.M.S. 12(1911), 75-98.

[45] A. Dold, "*Halbexakte Homotopiefunktoren,*" Lecture Notes in Mathematics 12, Springer-Verlag (1966).

[46] ———, "Über der Steenrodschen Kohomologie-operationen," Ann. of Math. 73(1961), 258-294.

[47] ———, "Homology of symmetric products and other functors of complexes," Ann. of Math. 68(1958), 54-80.

[48] A. Dold and R. Lashof, "Principal quasifibrations and fibre homotopy equivalence of bundles," Ill. J. Math. 3(1959), 285-305.

[49] A. Dold and R. Thom, "Quasifaserungen und unendliche symmetrische Produkte," Ann. of Math. 67(1958), 239-281.

[50] E. Dyer and R. Lashof, "Homology of iterated loop spaces," Amer. J. Math. 84(1962), 35-88.

[51] V.K.A.M. Gugenheim and J. P. May, "*On the Theory and Applications of Differential Torsion Products,*" Memoirs Amer. Math. Soc. 142 (1974).

[52] A. Hattori, "Integral characteristic numbers for weakly almost complex manifolds," Topology 5(1966), 259-280.

[53] M. Hirsch and B. Mazur, "*Smoothings of piecewise linear manifolds,*" Ann. of Math. Studies 80, Princeton 1974.

[54] F. Hirzebruch, "*Topological Methods in Algebraic Geometry,*" Springer-Verlag (1966).

[55] L. Hodgkin, "The K-theory of some well-known spaces I: $Q(S^0)$," Topology 11(1972), 371-375.

[56] L. Hodgkin and V. Snaith, "*Topics in K-theory, Two Independent Contributions,*" Lecture Notes in Mathematics 496, Springer-Verlag, Berlin (1975).

[57] J. F. P. Hudson, "*Piecewise linear topology,*" W. A. Benjamin, (1969).

[58] D. Husemoller, "The structure of the Hopf algebra $H_*(BU)$ over a $Z_{(p)}$-algebra," Amer. J. Math. 93(1971), 329-349.

[59] D. Husemoller, "*Fibre Bundles*," McGraw-Hill 1966.

[60] I. M. James, "Reduced product spaces," Ann. of Math. 62 (1955), 170-197.

[61] M. A. Kervaire, J. W. Milnor, "Groups of homotopy spheres I," Ann. of Math. 77 (1963), 504-537.

[62] ________, "Bernoulli numbers, homotopy groups and a theorem of Rohlin," Proc. Int. Congress of Math., Edinborough, 1958.

[63] R. Kirby, "Lectures on triangulations of Manifolds," mimeographed notes, UCLA (1969).

[64] R. Kirby, L. Siebenmann, "Some theorems on topological manifolds," *Manifolds - Amsterdam 1970*, Lecture Notes in Mathematics 197, Springer-Verlag (1971), 1-7.

[65] ________, *Foundational Essays on Topological Manifolds, Smoothings and Triangulations*. Ann. of Math. Studies 88, Princeton Press (1977).

[66] J. M. Kister, "Microbundles are fibre bundles," Ann. of Math. 80 (1964), 190-199.

[67] D. Kraines, "Massey higher products," Trans AMS. 124 (1966), 431-449.

[68] N. H. Kuiper, R. Lashof, "Microbundles and bundles I: Elementary theory," Invent. Math. 1 (1966), 1-17.

[69] R. Lashof, M. Rothenberg, "Microbundles and smoothing," Topology 3 (1965), 357-388.

[70] ________, "On the Hauptvermutung, triangulation of manifolds and h-cobordism," Bull. A.M.S. 72 (1966), 1040-1043.

[71] R. Lashof, "Poincaré duality and cobordism," Trans. A.M.S. 109 (1963), 257-277.

[72] H. Ligaard, I. Madsen, "Homology operations in the Eilenberg-Moore spectral sequences," Math. Z. 143 (1975), 45-54.

[73] I. Madsen, "On the action of the Dyer-Lashof algebra in $H_*(G)$," Pacific J. Math. 69 (1975), 235-275.

[74] ________, "Higher torsion in SG and BSG," Math. Z. 143 (1975), 55-80.

[75] I. Madsen, "The surgery formula and homology operations," Proceedings of the Advanced study institute on algebraic topology (Aarhus, 1970).

[76] ———, "Homology operations in $H_*(G/TOP)$," (to appear).

[77] ———, "Remarks on normal invariants from the infinite loop space viewpoint," *Algebraic and Geometric Topology*, Proc. A.M.S. Symposia in Pure Math. 32, A.M.S. (1978), 91-102.

[78] I. Madsen, R. J. Milgram, "The universal smooth surgery class," Comm. Math. Helv. 50 (1975), 281-310.

[79] I. Madsen, J. Tornehave, V. Snaith, "Infinite loop maps in geometric topology," Math. Proc. Camb. Phil. Soc. 81 (1977), 399-430.

[80] S. MacLane, "*Homology*," Springer-Verlag (1963).

[81] B. Mann, "The cohomology of the symmetric group," Dissertation, Stanford University, 1975.

[82] J. P. May, "Matrix Massey products," J. Algebra 12 (1969), 533-568.

[83] ———, "The algebraic Eilenberg-Moore spectral sequence," Preprint, University of Chicago.

[84] ———, "*The geometry of iterated loop spaces*," Lecture Notes in Mathematics, No. 271, Springer-Verlag (1972).

[85] J. P. May (with contributions by F. Quinn, N. Ray and J. Tornehave), "*E_∞-ring spaces and E_∞-ring spectra*," Lecture Notes in Mathematics, No. 577, Springer-Verlag (1977).

[86] F. R. Cohen, T. J. Lada, J. P. May, "*The homology of iterated loop spaces*," Lecture Notes in Mathematics, 533 Springer-Verlag (1976).

[87] W. Massey, "Higher order linking numbers," Conference on Algebraic Topology, University of Ill. at Chicago Circle (1968).

[88] R. J. Milgram, "The bar construction and abelian H-spaces," Ill. J. Math. 11 (1967), 242-250.

[89] ———, "Steenrod squares and higher Massey products," Bull. Soc. Mat. Mex. (1968), 32-51.

[90] ———, "The mod. 2 spherical characteristic classes," Ann. of Math. 92 (1970), 238-261.

[91] R. J. Milgram, "Surgery with coefficients," Ann. of Math. 100 (1974), 194-265.

[92] ________, "The homology of symmetric products," Trans. A.M.S. 138 (1969), 251-265.

[93] ________, "The structure over the Steenrod algebra of some 2-stage Postnikov systems," Quart. J. Math. 20 (1969), 161-169.

[94] J. Milnor, "On axiomatic homology theory," Pacific J. Math. 12 (1962), 337-341.

[95] ________, "On the cobordism ring Ω_* and a complex analogue, Part I," Amer. J. Math. 82 (1960), 505-521.

[96] ________, "Microbundles I," Topology 3, suppl. 1 (1964), 53-81.

[97] ________, "*Lectures on the* h-*cobordism theorem*," Princeton University Press, Princeton 1965.

[98] ________, "Construction of universal bundles II," Ann. of Math. 63 (1956), 430-436.

[99] ________, "*Singular points of complex hypersurfaces*," Ann. of Math. studies 61 (1968).

[100] ________, "On spaces having the homotopy type of a CW-comples," Trans. A.M.S. 90 (1959), 272-280.

[101] J. Milnor, J. Moore, "On the structure of Hopf algebras," Ann. of Math. 81 (1965), 211-264.

[102] J. Milnor, J. Stasheff, "*Characteristic classes*," Ann. of Math. studies 76 (1974).

[103] J. Morgan, D. Sullivan, "The transversality characteristic class and linking cycles in surgery theory," Ann. of Math. 99 (1974), 384-463.

[104] E. L. Moore, "A two-fold generalization of Fermat's theorem," Bull. A.M.S. 2nd series 2 (1896), 189-199.

[105] R. J. Milgram, L. Moser, "A note on the Kummer surface," (mimeographed), Hayward State University, Calif.

[106] H. Mui, "Modular invariant theory and cohomology algebras of symmetric groups," J. Fac. Sci. Tokyo 22 (1976), 319-371.

[107] M. Nakaoka, "Homology of the infinite symmetric group," Ann. of Math. 73 (1961), 229-257.

[108] G. Nishida, "Cohomology operations in iterated loop spaces," Proc. J. Acad. 44 (1968), 104-109.

[109] S. P. Novikov, "Homotopy equivalent smooth manifolds I," Translations A.M.S. 48 (1965), 271-396.

[110] F. Peterson, "Some results on PL-cobordism," J. Math. Kyoto Univ. 9 (1969), 189-194.

[111] D. Quillen, "The Adams conjecture," Topology 10 (1971), 67-80.

[112] ———, "On the completion of a simplicial monoid," (Preprint), M.I.T.

[113] F. Quinn, "Surgery on Poincaré and normal spaces," mimeographed notes, New York University (1971).

[114] D. Ravenel, W. Wilson, "Bipolynomial Hopf algebras," J. Pure and Applied Algebra 4 (1974), 41-45.

[115] M. Rothenberg, N. E. Steenrod, "The cohomology of classifying spaces of H-spaces," Bull. A.M.S. 71 (1965), 872-875.

[116] C. Rourke, B. Sanderson, "*Introduction to Piecewise-linear Topology*," Ergebnisse 69, Springer-Verlag (1972).

[117] C. Rourke, D. Sullivan, "On the Kervaire obstruction," Ann. of Math. 94 (1971), 397-413.

[118] G. B. Segal, "The representation ring of a compact Lie group," Publ. Math. I.H.E.S. 39 (1968), 113-128.

[119] ———, "Classifying spaces and spectral sequences," *ibid.* Publ. Math. I.H.E.S. 34 (1968), 105-112.

[120] ———, "Categories and cohomology theories," Topology 13 (1974), 293-312.

[121] ———, "Configuration-spaces and iterated loop spaces," Invent. Math. 21 (1973), 213-221.

[122] J. P. Serre, "Groupes d'homotopie et classes de groupes abéliens," Ann. of Math. 58 (1953), 258-294.

[123] J. P. Serre, "Cohomologie modulo 2 des complexes d'Eilenberg-MacLane," Comm. Math. Helv. 27 (1953), 198-232.

[124] S. Smale, "Generalized Poincaré conjecture in dimensions greater than 4," Ann. of Math. 74 (1961), 391-406.

[125] E. H. Spanier, "*Algebraic Topology,*" McGraw Hill (1966).

[126] ———, "The homology of Kummer Manifolds," Proc. A.M.S. 7 (1956), 155-160.

[127] M. Spivak, "Spaces satisfying Poincaré duality," Topology 6 (1969), 77-102.

[128] J. Stasheff, "A classification theorem for fiber spaces," Topology 2 (1963), 239-246.

[129] N. E. Steenrod, "*Topology of Fibre Bundles,*" Princeton University Press (1951).

[130] N. E. Steenrod, D. Epstein, "*Cohomology Operations,*" Annals of Math. Studies 50, Princeton (1962).

[131] N. E. Steenrod, "Milgrams classifying space of a topological group," Topology 7 (1968), 349-368.

[132] ———, "*Cohomology Operations and Obstructions to Extending Continuous Functions,*" A.M.S. Colloquium Lectures (1957). (See also Seminaire H. Cartan (Exposé 3) Secrétariat Math., Paris, 1954-1955.)

[133] R. E. Stong, "*Notes on Cobordism Theory,*" Princeton University Press (1968).

[134] D. Sullivan, Thesis, Princeton University (1965).

[135] ———, "Triangulating homotopy equivalences," mimeographed notes, Warwick University (1965).

[136] ———, "Genetics of homotopy theory and the Adams conjecture," Ann. of Math. 100 (1974), 1-80.

[137] R. Thom, "Quelques propriétés globales des variétés differentiables," Comm. Math. Helv. 28 (1954), 17-86.

[138] ———, "Les classes characteristiques de Pontrjagin des variétés triangulées," Sym. Int. de Topologia Algebraica, Mexico (1958).

[139] A. Tsuchiya, "Characteristic classes for spherical fiber spaces," Nagoya Math. J. 43 (1971), 1-39.

[140] ———, "Characteristic classes for PL microbundles," Nagoya Math. J. 43 (1971), 169-198.

[141] C. T. C. Wall, "*Surgery on Compact Manifolds*," Academic Press, (1970).

[142] ———, "Determination of the cobordism ring," Ann. of Math. 72 (1960), 292-311.

[143] ———, "Cobordism of combinatorial n-manifolds for $n \leq 8$," Proc. Cambridge Phil. Soc. 60 (1964), 807-811.

[144] R. Woolfson, *Hyper Γ-spaces and Γ-structures*, (Thesis, Oxford 1976).

[145] G. W. Whitehead, "On the homotopy groups of spheres and rotation groups," Ann. of Math. 43 (1942), 634-640.

[146] R. E. Williamson, "Cobordism of combinatorial manifolds," Ann. of Math. 83 (1966), 1-33.

[147] Z. I. Yosimura, "Universal coefficient sequence for cohomology theories of CW spectra," Osaka J. Math. 12 (1975), 305-323.

[148] D. Anderson, Thesis, University of California, Berkeley (1964).

[149] M. F. Atiyah, "Characters and Cohomology of finite groups," Publ. Math. I.H.E.S., 9 (1961), 247-288.

[150] A. Clark, "Homotopy commutativity and the Moore spectral sequence," Pacific J. Math. 15 (1965), 65-74.

[151] S. Eilenberg, J. Moore, "Homology and fibrations," Comm. Math. Helv. 40 (1966), 199-236.

[152] L. Jones, "Patch spaces: A geometric representation for Poincaré spaces," Ann. of Math. 97 (1973), 276-306.

[153] L. Kristensen, "On the cohomology of spaces with two non-vanishing homotopy groups," Math. Scand. 12 (1963), 83-105.

[154] A. Liulevicius, "Representation rings of symmetric groups – a Hopf algebra approach," (to appear).

[155] I. Madsen, R. J. Milgram, "On spherical fibre bundles and their PL reductions," *Recent developments in topology*, (ed. G. Segal), Oxford (1973).

[156] J. Milnor, "The Steenrod algebra and its dual," Ann. of Math. 67 (1958), 150-171.

[157] J. P. May, "A general algebraic approach to Steenrod operations," Springer Lecture Notes in Mathematics, 168 Springer-Verlag (1970), 153-231.

[158] B. Mann, R. J. Milgram, "On the action of the Steenrod Algebra $\mathcal{A}(p)$ on $H^*(MSPL, Z/p)$ at odd primes," Mimeo. Stanford Univ. (1976).

[159] H. Ligaard, J. P. May, "On the Adams spectral sequence for $\pi_*(MSTOP)$, I," Mimeo. University of Chicago (1976).

[160] H. Ligaard, B. Mann, J. P. May, R. J. Milgram (to appear).

INDEX

α , number of ones in dyadic expansion, 54, 196

α_p , solution to Adams conjecture at p, 106

Adams conjecture, 106

Adem relations, 129

admissible, 142-143

$\mathfrak{A}$-genus, 104-105

almost complex manifold, 20

Anderson splitting of $BSO^{\oplus}[p]$, 104

Arf invariant, 37

Atiyah-Bott-Shapiro orientation of Spin-bundles, 108

Atiyah-Hirzebruch spectral sequence, 86, 91, 160-161

Atiyah-Segal exponential isomorphism 104, 182

Atiyah-Segal, KO-theory of BG, 94

Aut(π), automorphism group of a group π, 51

Aut(X), group of homotopy classes of homotopy equivalences of X, 41

β , Bockstein operator, 96

β_p , splitting map $G/O[p] \to BSO[p]$, 107-108

B_n , n'th Bernoulli number, 209-210

Barratt-Priddy-Quillen theorem, 50

$B\operatorname{cok} J_p$, 102

Becker-Gottlieb transfer, 120-121

Bernoulli numbers, 117, 167, 186, 209-210

Bernoulli polynomials, 213

BG, classifying space for spherical fibrations, 45

bipolynomial Hopf algebra, 180

BO, classifying space for real vector bundles, 8
- cohomology and homology mod 2, 13
- integral cohomology and homology, 13, 235
- BO[½], 23

Bockstein spectral sequence, 146

BΠ, classifying space for Π-bundles, 4
- construction, 5-7

BPL, classifying space for PL $\mathbf{R}^n$-bundles, *see* BTOP

Brieskorn varieties, 167, 173

Browder-Novikov theorem, 37

BSG, classifying space for oriented spherical fibrations, 45
- cohomology mod 2, 70
- Pontrjagin ring mod 2, 139

BSO, classifying space for oriented real vector bundles, 8
- cohomology mod 2, 14
- KO-theory, 94
- splitting at p, 104
- free integral cohomology, 185

$BSO_{(1)}$, $BSO^{\perp}_{(1)}$, 105

BSpin, classifying space for Spin-bundles, 108

BTOP, classifying space for $\mathbf{R}$-bundles (with 0-section), 9
- free integral homology and cohomology, 197-200
- generators for free homology at 2, 205

– generators for free homology away from 2, 215
– homology mod 2, 232, 234
– higher 2-torsion, 238-245
BU, classifying space for complex vector bundles, 8
– integral homology and cohomology, 11, 12
(*see also* $H_d(\mathbf{Z})$)

χ, canonical antiautomorphism, 67
c_i, i'th Chern class, 10-11
Cannibalistic class ρ^k, 102
Cerf theorem on Diff(S^3), 33, 43
ch, Chern character, 84
characteristic power series, 24-25
Chern character, 84
Chern class, 10-11
classifying space, 3
cobordism, 16
cok J_p, 108-109
connected sum (along boundary) #, 39
Conner-Floyd theorem, 85

δ, transformation $\Omega_* \to KO_*[½]$, 84-85, 100
Δ, orientation of a Thom spectrum, 82
Δ_A, BO-orientation of MSpin, 108
Δ_{PL}, BO[½]-orientation of MSPL, 100
Δ_{SO}, BO[½]-orientation of MSO, 84
d_2, structure map for infinite loop spaces, 126-127
$\hat{d}_2$, multiplicative structure map for QS^0, 130
d_n, structure map for infinite loop spaces, 119
detecting family of subgroups, 54
divided power algebra, 189
Dold-Thom theorem, 62
Dyer-Lashof algebra, 132
Dyer-Lashof operations, 127

η, map $\Omega_*(G/PL) \to \Omega_*^{PL}$, 159
e_A, characteristic class for fiber homotopy trivial bundles, 108
e_L, characteristic class for fiber homotopy trivial bundles, 107
E, spectrum, 18
$E(\phi)$, equalizer, 191
$E_{(i_1,\cdots,i_n)}$, generator for $H_*(QS^0; \mathbf{Z}/2)$, 61-62, 64
$E_{n,m}$, divided power generator for $\Omega^{PL}/\text{Tor} \otimes \mathbf{Z}_{(2)}$, 207
Eilenberg-Maclane spectrum, 20, 82
Eilenberg-Moore spectral sequence, 7, 155
E-orientation, 82
$E\pi$, universal free π-space, 3
equalizer, 191
excess, 142
exotic complex projective space, 163, 167, 173

$\hat{\phi}_{4i+1}$, primitive classes in $H_*(B(G/O); \mathbf{Z}_{(2)})$, 26-27
F_n, based homotopy equivalences of S^n, 46
(F, Π)-bundle, 3
$F_*(X; A)$, free homology tensored with A, 24
$F^*(X; A)$, free cohomology tensored with A, 24

γ_n, n'th divided power, 189-190
γ_H^n, universal n-plane bundle over BH_n, 15
γ_p, mapping $BSO^{\oplus}[p] \to BSPL[p]$, 105
γ-operations, 181
Γ, divided power algebra, 189
$\Gamma \wr H$, wreath product of groups Γ, H, 53
Γ_n, quadratic construction, 125
– higher analogue in cohomology, 70-71

G_n, homotopy equivalences of S^{n-1}, 8-9, 45
generalized (co)homology theory, 18
genus, 24-25
G/O, classifying space for smooth normal invariants, 41
– Sullivan splitting, 108
– homology mod 2, 224
G/PL, classifying space for PL normal invariants, 41-42
– homotopy groups, 43-44
– homotopy type at 2 and away from 2, 79, 93
– H-space type, 97
– homotopy type of first two de-loopings at 2, 156-157
– free cohomology at 2 and away from 2, 187-188
– free homology, 189, 191-192
Group completion, 12, 126
G/TOP, 44
– homotopy groups, 44
– homotopy type, 98-99
– homotopy type of first two deloopings at 2, 41
– homology operations Q_0, Q_1, Q_2, 144-145
– free cohomology at 2 and away from 2, 188-189
– free homology, 189, 192

h, (generalized) Hurewicz map, 26-27, 77
$h_{n,i}$, Witt rector basis for $H_d(Z_{(p)})$, 178
$H_d(A)$, model Hopf algebra, 174
– automorphisms of $H_d(Z)$, 176
– dual, 176-177
– splitting theorem, 178
Hattori-Stong theorem, 218
– PL version, 220
h-cobordism theorem, 38
Hirzebruch index theorem, 26
– PL version, 88
Hodgkin-Snaith theorem, 115
homology operations, 127
homotopy F-bundle, 5
homotopy smoothing, 40
homotopy triangulation, 40
Hurewicz map, 26, 77
Husemoller splitting theorem, 178

i_n, inclusion $B\Sigma_n \to Q(S^0)$, 50
index, 26, 108
Ind_G^H, induction homomorphism, 122
induction, 122
infinite loop space, 21, 119, 124
Int(Π), group of inner automorphisms of Π, 52
inverse limit, 90
IO(G), augmentation ideal of RO(G), 94

j, map $BO \times G/TOP \to BTOP$, 194
$J_{n,m}$, homomorphism $\Sigma_n \wr \Sigma_m \to \Sigma_{nm}$, 53
$\hat{J}_{2,m}$, homomorphism $\Sigma_2 \wr \Sigma_m \to \Sigma_{m^2}$, 133
J_p, $J_p^{\otimes}$, 109-110
J-homomorphism, 47

k_{4n}, primitive fundamental class in $H^*(G/PL; Z_{(2)})$, 95
k_{4i+1}, primitive fundamental class in $H^*(B(G/TOP); Z_{(2)})$, 228
$\overline{K}_{4n}$, $\tilde{K}_{4n}$, fundamental class in $H^*(G/PL; Z_{(2)})$, 78, 96
K_{4n-2}, primitive fundamental class in $H^*(G/PL; Z/2)$, 78
K(A), Eilenberg-MacLane spectrum, 20, 82

$K(A, i)$, Eilenberg-MacLane space, 20
Kervaire manifold, 37, 165-166
$KO[½]$, real K-theory localized away from 2, 136, 82
– universal coefficient theorem, 85
Kummer congruences, 187
Kummer surface, 38, 173
$K(Z/2, n)$ and $K(Z_{(2)}, n)$, Eilenberg-MacLane spaces, 20
– cohomology mod 2, 143
– homology operations, 143
– Bockstein spectral sequence, 146

λ-operations, 181
$\mathcal{L}$-genus, 26
$\mathcal{L}_{PL}$-genus, 88
$\varprojlim$, inverse limit, 90
$\varprojlim^{(1)}$, derived functor of $\varprojlim$, 90
localization, 21
localized at p, 22
localized away from p, 22
loop space, 46

M_A^{2n}, Kervaire manifold, 165
M_B^{2n}, index 8 Milnor manifold, 166
Massey product, 149
matric Massey product, 151
May formula, 135
May-Kraines suspension theorem, 153
$M\operatorname{cok} J_p$, Thom spectrum for $B\operatorname{cok} J_p$, 113, 116
$\mathfrak{M}$-genus, 187-188
MH, Thom spectrum for H-bundles, 18-19
Milnor criteria for indecomposables in $\Omega_*^{SO}/\mathrm{Tor}$, 221
Milnor $\varprojlim^{(1)}$-sequence, 91
Milnor manifold, 37, 166
– Pontrjagin numbers, 200-201
mixed Adem relation,
mixed Cartan formula, 132
MO, Thom spectrum for real vector bundles, 20
Moore space, 21
MSO, Thom spectrum for oriented real vector bundles, 23-24
MSPL, Thom spectrum for oriented PL-bundles, 20
– Sullivan splitting, 114
multiplicative characteristic class, 24
multiplicative E-orientation, 83

ν_2, 2-adic valuation, 54
$\mathfrak{N}_*$, unoriented smooth cobordism, 19-20
Nishida relations, 129
$N_m(n, k)$, 206
$NM_0(X)$, smooth normal invariants of X, 36-37, 42
$NM_{PL}(X)$, PL normal invariants of X, 36-37, 42
normal cobordism, 36
normal invariant, 34
normal map, 35
Novikov classification theorem, 39
$\mathfrak{N}_*^{PL}$, unoriented piecewise linear cobordism, 19-20, 252
$\mathfrak{N}_*^{TOP}$, unoriented topological cobordism, 19-20, 223, 252-253

ΩX, loop space of X, 46
Ω_*, oriented smooth cobordism, 19, 23

$\Omega^n X$, n'th loop space of X, 47
$\Omega^n_\alpha X$, path component of $\Omega^n X$, 47
Ω^{PL}_*, oriented piecewise linear cobordism, 19-20
– $\Omega^{PL}_*/\mathrm{Tor}$, set of generators, 163, 173
– $\Omega^{PL}_*/\mathrm{Tor}$, 2-local structure, 206, 207
– $\Omega^{PL}_*/\mathrm{Tor}$, p-local structure, 219
– criteria for indecomposability, 222
– 2-torsion, 246-253
Ω^{TOP}_*, oriented topological cobordism, *see* Ω^{PL}_*

$\psi_{n,m}$, homomorphism $\Sigma_n \times \Sigma_m \to \Sigma_{nm}$, 50
p_{4i}, Pontrjagin class in degree 4i, 88-89
P_n, simply connected surgery obstruction group, 40
ph, Pontrjagin character, 84
phantom map, 91
piecewise linear Pontrjagin classes, 88-89
PL_n, simplicial group of piecewise linear homomorphisms of $\mathbf{R}^n$, 9
PL/O classifying space for smoothings, 33, 88-89
plumbing, 164
Poincaré duality space, 29
polyhedral path lifting property, 46
Pontrjagin character, 84-85
Pontrjagin classes, 13, 88, 217
Pontrjagin-Thom map, 19, 30-31
primitive generators for $H^{4n}(G/PL;\mathbf{Z}_{(2)})$, 96
primitive series, 25-26
principal Π-bundle, 3-4

Q, free infinite loop space functor, 48-49
Q^i, Dyer-Lashof operation, 128
Q_j, Dyer-Lashof operation, 127
$\hat{Q}_j$, multiplicative Dyer-Lashof operation, 132
$Q(S^0)$, stable self maps of spheres, 48
– homology mod 2, 64, 137
– cohomology mod 2, 65
quadratic construction, 54

ρ^k_A, mapping $BSO^{\oplus}[p] \to BSO^{\otimes}[p]$, 103
ρ^k_L, mapping $BSPL \to BO^{\otimes}[\frac{1}{2}, \frac{1}{k}]$, 102
$\mathcal{R}$, Dyer-Lashof algebra, 138-139
Ravenel-Wilson theorem, 180
$\mathcal{R}$-genus, 209
ring spectrum, 82
RO(G), real representation ring of G, 94

Σ_n, symmetric group of degree n, 49
– homology mod 2, 63
s_I, index surgery obstruction, 76
s_K, Kervaire surgery obstruction, 76
s_n, Newton polynomial, 174
$\mathcal{S}_0(X)$, homotopy smoothings of X, 40
$\mathcal{S}_{PL}(X)$, homotopy triangulations of X, 40
S-duality, 31-32
Serre fibration, 46
SF_n, oriented based homotopy equivalences of S^n, 47
SG, stable homotopy equivalences of spheres, 45

– homology mod 2, 54, 138
– Sullivan splitting, 110
SG_n, oriented homotopy equivalences of S^{n-1}, 45
singular manifold, 16
Spanier-Whitehead duality, 32
spectrum, 18
spherical fibration, 9, 30
Spivak normal bundle, 30
SP^n, n-fold symmetric product functor, 62
Sq^n, n'th Steenrod square, 56
Steenrod recognition principle, 4
Steenrod squares, 56
Stiefel-Whitney classes, 10, 59
structure maps for infinite loop spaces, 119, 126
Sullivan orientation of PL-bundles away from 2, 99
Sullivan splitting of BSPL, 105
Sullivan splitting of G/O and SG, 110
Sullivan splitting of MSPL, 114
Sullivan's analysis of G/PL, 79, 93, 97
Sullivan's surgery formula, 80
surgery exact sequence, 40
surgery problem, 35
surgery obstruction groups, 40

θ, mapping $BSO \times G/PL \to BSPL$, 159
T_{4n}, generators of $F^*(BTOP;\mathbf{Z}_{(2)})$, 205
Thom class, 27, 30-31
Thom isomorphism, 31, 83
Thom space, spectrum, 18, 30, 113
TOP_n, based homeomorphisms of $\mathbf{R}^n$, 8
TOP/PL, 44
transfer, 53, 20
transversality, 17

u, retraction $Q(X) \to X$ of infinite loop spaces, 125
U, Thom class, 27, 31
U_{PL}, Thom class, 102
unit of a spectrum, 82

V, total Wu class, 80
V_n, elementary abelian subgroup of Σ_{2^n}, 56

w_i, Stiefel-Whitney class, 10-11
Witt vector basis, 178
$W_{d,n}(\mathbf{Z}_{(p)})$, 178
Wreath product, 53
Wu-class, 80
Wu-formula, 59-60

X[p], localization of X at p, 22
X[1/p], localization of X away from p, 22
X[Q], rational localization of X, 23

$\mathbf{Z}_{(p)}$, integers localized at p, 22
$\mathbf{Z}[1/p]$, integers localized away from p, 22

Library of Congress Cataloging in Publication Data

Madsen, Ib.
The classifying spaces for surgery and cobordism of manifolds.

(Annals of mathematics studies ; no. 92)
Bibliography: p.
Includes index.
1. Manifolds (mathematics) 2. Classifying spaces. 3. Surgery (Topology) 4. Cobordism theory.
I. Milgram, R. James, 1939- joint author.
II. Title. III. Series.
QA613.2.M32 514'.223 78-70311
ISBN 0-691-08225-1
ISBN 0-691-08226-X pbk.